CSR, Sustainability, Ethics & Governance

Series Editors
Samuel O. Idowu
London Metropolitan University, Calcutta House
London, UK
René Schmidpeter
BFH, Bern, Switzerland

In recent years the discussion concerning the relation between business and society has made immense strides. This has in turn led to a broad academic and practical discussion on innovative management concepts, such as Corporate Social Responsibility, Corporate Governance and Sustainability Management. This series offers a comprehensive overview of the latest theoretical and empirical research and provides sound concepts for sustainable business strategies. In order to do so, it combines the insights of leading researchers and thinkers in the fields of management theory and the social sciences and from all over the world, thus contributing to the interdisciplinary and intercultural discussion on the role of business in society. The underlying intention of this series is to help solve the world's most challenging problems by developing new management concepts that create value for business and society alike. In order to support those managers, researchers and students who are pursuing sustainable business approaches for our common future, the series offers them access to cutting-edge management approaches. CSR, Sustainability, Ethics and Governance is accepted by the Norwegian Register for Scientific Journals, Series and Publishers, maintained and operated by the Norwegian Social Science Data Services (NSD).

Manuela Gussoni

Integrating Strategic Sustainability Practices in Business

A Guide for Sustainability Consultants

Springer

Manuela Gussoni
Bureau Veritas Nexta
Rome, Italy

ISSN 2196-7075 ISSN 2196-7083 (electronic)
CSR, Sustainability, Ethics & Governance
ISBN 978-3-032-09616-6 ISBN 978-3-032-09617-3 (eBook)
https://doi.org/10.1007/978-3-032-09617-3

This Springer imprint is published by the registered company Springer Nature Switzerland AG
The registered company address is: Gewerbestrasse 11, 6330 Cham, Switzerland

Contents

1 Strategic Consulting in the Sustainability Era

Strategic consulting represents a multidisciplinary approach that combines diverse skill sets and specialized vertical competencies to guide organizations in making informed investment decisions and achieving sustainable long-term objectives. Strategic consultants are typically engaged in high-level decision-making processes to ensure that all dimensions of a business challenge are thoroughly considered. Their value lies in the ability to synthesize analytical, interpersonal, and industry-specific expertise to co-design tailored strategies and solutions that align with the client's strategic vision.

According to industry experts, the strategic consulting sector accounts for 32% of the total consulting market[1] and it is organized in different verticals such as Mergers and Acquisitions, Business Model Transformation, Economic Policy, Organizational Strategy, and Corporate Strategy.

[1] Fortune Business Insights. (2024). Strategic Consulting Market Size, Share & Industry Analysis, By Service Type, Enterprise Type, Industry, and Regional Forecast, 2025–2032.

M. Gussoni, *Integrating Strategic Sustainability Practices in Business*, CSR, Sustainability, Ethics & Governance,
https://doi.org/10.1007/978-3-032-09617-3_1

Focus ON: The Consulting Industry in a Nutshell

The consulting industry can trace its roots back to the late nineteenth century; however, it was not until the 1930s that consulting firms started to grow their size. The consulting industry started its internationalization in the early 1960s, when the large American management consultancies expanded in Europe, and in the next 30 years, the consulting industry found itself in a phase of unprecedented growth.[2] In particular, it is worth to underline the dramatic growth of the industry between 2020 and 2023 (about 9% CAGR) (Source: Statista) thanks to the resilience demonstrated, during the pandemic, shifting to a remote delivering model and some significant trends like digital transformation and a heightened focus on sustainability.[3]

In 2023, the consulting industry market size is an estimated $1000B (Source: Statista). These firms range from specialized boutique consulting firms to the Big 4 that employ hundreds of thousands of people.[4] Big 4 companies are: Deloitte, PwC, Ernst & Young, and KPMG. They account for about 20% of the total industry Revenues.[5]

Deloitte is the largest big 4 consulting firm; it offers a comprehensive suite of services, including audit, consulting, financial advisory, risk advisory, and tax services. The firm is known for its expertise in areas such as digital transformation, strategy, and risk management, helping clients navigate complex business challenges. Deloitte positions sustainability as a driver of long-term business transformation and measurable impact. Their approach emphasizes embedding sustainability into every facet of operations, turning challenges into growth opportunities, and aligning with net-zero goals.[6] It provides end-to-end sustainability solutions, i.e. services that cover the entire lifecycle of sustainability strategy and implementation within an organization.

PwC provides a wide range of services, including audit, assurance, advisory, and tax services. The firm is particularly renowned for its expertise in areas such as cybersecurity, data analytics, and sustainability. Among sustainability services,[7] it offers a holistic approach to help business leaders in facing a complex landscape—shifting regulations, changing tax and tariffs, rising energy costs, and supply chains exposed to climate risks.

---

[2] https://www.consultancy.eu/consulting-industry

[3] Management Consultancies Association. (2024). *MCA Member Survey 2023*. https://www.mca.org.uk/report/mca-member-survey-2023

[4] https://managementconsulted.com/revenue-of-top-20-firms/

[5] Estimation on the basis of companies' annual reports.

[6] https://www.deloitte.com/global/en/issues/climate/sustainability-and-climate.html?icid=top_sustainability-and-climate

[7] https://www.pwc.com/gx/en/services/sustainability.html

EY offers a diverse range of services, including assurance, consulting, strategy and transactions, and tax services. The firm is known for its innovative approach, leveraging emerging technologies and data-driven insights to help clients achieve their strategic objectives. Even EY offers a comprehensive and structured portfolio of sustainability services designed to help organizations integrate environmental, social, and governance (ESG) principles into their core strategies and operations.[8] EY's value proposition centers on creating long-term stakeholder value, building trust in capital markets, empowering clients through data analytics and technological innovation.

Finally, KPMG provides audit, advisory, and tax services, with a focus on areas such as digital transformation, risk management, and compliance. The firm is known for its industry-specific expertise, helping clients navigate complex regulatory environments and optimize their operations. Also KPMG recognizes the critical role sustainability plays in driving long-term business success[9] and offers a wide range of services like ESG strategy development, climate risk and decarbonization planning, sustainable finance advisory, and support for regulatory compliance and reporting. KPMG also assists clients in integrating sustainability into governance structures, supply chains, and digital transformation initiatives.

In conclusion, all of the Big Four consulting firms—Deloitte, EY, PwC, and KPMG—have strategically integrated sustainability into their core service offerings, recognizing it is a fundamental driver of long-term growth, resilience, and value for their clients. Moreover, it is also worth mentioning three additional prestigious management consulting firms—McKinsey & Company, Boston Consulting Group (BCG), and Bain & Company—all of which place sustainability at the core of strategic planning.

It is evident how, today, sustainability issues become central in corporate strategic decision-making and taking into account Environmental, Social, and Governance (ESG) factors is an essential prerequisite for a long-term business success.

At this regard, some studies have demonstrated that an effective ESG Risks Management by a company is correlated with a higher profitability (Giese et al., 2019[10]), a better resources management (Gregory et al., 2014[11]), and the reduction

[8] https://www.ey.com/en_gl/services/sustainability/services

[9] https://kpmg.com/us/en/kpmg-sustainability.html

[10] Giese G., Lee L.E., Melas D., Nagy Z. E and Nishikawa L. (2019), "Foundation of ESG Investing: How ESG Affects Equity Valuation, Risk and Performance", The Journal of Portfolio Management, 45(5), 1–15

[11] Gregory A., Tharyan R. e Whittaker J. (2014), "Corporate Social Responsibility and Firm Value: Disaggregating the Effects on Cash Flow, Risk and Growth", Journal of Business Ethics, 124 (4), 633–657

of the specific risk component (Chen et al., 2018[12]) with a direct consequence on the capital cost. The less the ESG risks, the more the corporate value on the market.

Given these premises, corporations are increasingly seeking strategic consultants who not only support the development and execution of competitive and cost-efficient strategic plans, but also ensure alignment with evolving regulatory frameworks and sustainability imperatives. Today, being compliant is no longer sufficient—organizations must also demonstrate environmental and social responsibility. Strategic consultants are therefore expected to help clients minimizing costs, maximizing return on investment, and simultaneously embed sustainability into their core strategies, enabling long-term growth and resilience.

The sustainable strategy consultant, however, represents a relatively new professional figure within the consulting industry, emerging in response to the growing demand for expertise in sustainability-driven transformation. This book seeks to provide a thorough exploration of the consultant's principal functions, responsibilities, and essential competencies. Furthermore, it offers a practical framework to support consultants in addressing sustainability-related strategic challenges through the development of customized advisory services.

This book will present, as well, how sustainability embraces business strategy and how the role of the sustainable strategic consultant becomes central in helping corporates in taking the best investment decisions to improve their sustainability rating, becoming more appealing for investors and, at the same time, compliant with the normative framework requirements. These professionals are, in fact, responsible for conducting environmental impact assessments, developing sustainability strategies, advising on regulatory compliance, and integrating sustainable practices across organizational functions.

For these reasons, this book is primarily intended for strategic consultants seeking to deepen their expertise in sustainability-related issues and it is aimed at enhancing their ability to deliver integrated advisory services to support organizations in transitioning toward sustainable business models—improving environmental and social performance while generating long-term value for stakeholders. Through actionable frameworks, real-world examples, and best practices, the book equips professionals to guide organizations in aligning their business models with environmental, social, and governance (ESG) principles—ensuring resilience, regulatory compliance, and competitive advantage in a rapidly evolving global landscape.

The book is structured as follows (See: Fig. 1.1):

Chapter 2 is a comprehensive presentation of the sustainable finance instruments and the international and the European normative framework in this field, with a focus on the recently approved Corporate Sustainability Reporting Directive (CSRD) and the Corporate Sustainability Due Diligence Directive (CSDDD).

[12] Chen Y.C., Hung M. e Wang H. (2018), "The Effect of mandatory CSR disclosure on firm profitability and social externalities: Evidence from China", Journal of Accounting and Economics, 65(1), 169–190

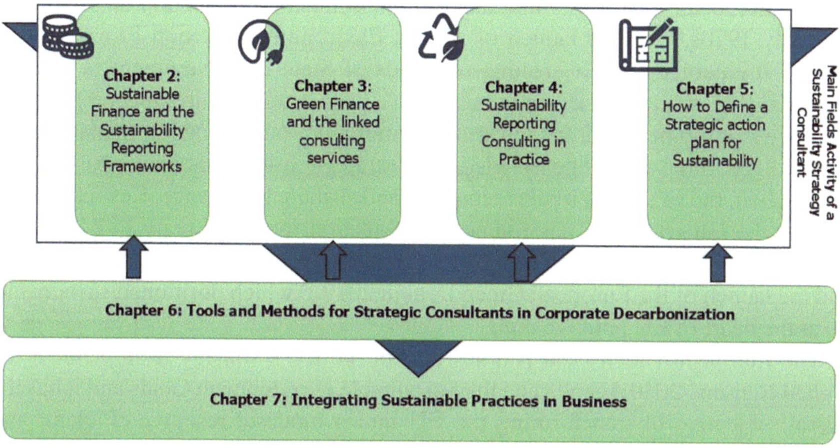

Fig. 1.1 The structure of the Book

The International Frameworks for ESG disclosure and sustainability reporting are presented with a deep focus, in particular, on the IFRS and GRI Standards.

Chapter 3 is an introduction to green finance and a presentation of the linked consulting services.

Chapter 4 is about Sustainability Assessment and Reporting and it will be explained in practice how a Sustainability Report has to be structured with the presentation of some case studies and examples.

Chapter 5 presents an effective approach to sustainable strategic planning from the analysis of the state of the art and alignment of company procedures to ESG management; to the definition of an action plan based on the company's weaknesses as identified through a benchmarking analysis against peers.

Chapter 6 will provide Sustainable Strategic Consultants with a toolkit to drive corporate decarbonization, including an introduction to LCA and GHG calculation methods.

Finally, Chap. 7, building on the previous Chapters, describes how to integrate sustainable practices in business, leading to improved risk management and an increased company's long-term value.

1.1 The European Strategy for Sustainability: From the European Green Deal to Date

Europe has long established itself as a global frontrunner in sustainability, distinguished by its pioneering policy frameworks, regulatory leadership, and consistent performance in advancing the United Nations Sustainable Development Goals (SDGs). Consultants operating in the European market are expected not only to navigate increasingly complex normative landscapes, but also to help clients align with ambitious environmental and social standards.

The European Union's commitment to environmental protection dates back to the early 1970s, with the launch of its first Environmental Action Programme in 1973.[13] It emerged from a meeting of Heads of State or Government in Paris in October 1972, who emphasized the need for a Community environmental policy.

Since that moment, European Union commitment toward environmental protection evolved into a comprehensive legal and strategic framework encompassing climate action, biodiversity, circular economy, and sustainable finance. Key milestones include the integration of environmental protection into the Treaty of Maastricht (1993),[14] the adoption of the European Sustainable Development Strategy (2001),[15] and the launch of the European Green Deal in 2019,[16] which positioned sustainability at the heart of EU policymaking.

The European Green Deal is a key part of the EU's strategy to implement the United Nations' 2030 Agenda and the Sustainable Development Goals and it has the overall objective of transforming the EU into a modern, resource-efficient, and competitive economy, with the ultimate goal of achieving climate neutrality by 2050. It also aims to protect, conserve, and enhance the EU's natural capital, and protect the health and well-being of citizens from environment-related risks and impacts, granting a just and inclusive transition (See: Fig. 1.2).

In order to reach the overall objective to achieve climate neutrality by 2050, the European Green Deal outlines a comprehensive **set of specific objectives** achievable by the implementation of deeply transformative policies across different sectors, building on the European strategic long-term vision set out in 2018 with the communication "A Clean Planet for all."[17]

The European Green Deal (EGD) specific objectives are:

1. **GHG Reduction**: The EGD wants to enhance the EU's greenhouse gas emission reductions target for 2030 to at least 50% and toward 55% compared with 1990 levels. In 2021, the European Climate Law (EU Regulation 2021/1119) set the objective of Climate Neutrality in the Union by 2050.

---

[13] Council of the European Communities. (1973). Declaration on the programme of action of the European Communities on the environment. Publications Office of the European Union. https://op.europa.eu/en/publication-detail/-/publication/f54da63e-be4d-4664-a8ea-aa678e640b51/language-en

[14] Treaty on European Union (Treaty of Maastricht), November 1, 1993. https://eur-lex.europa.eu/legal-content/EN/TXT/?uri=CELEX%3A11992M%2FTXT

[15] European Council. (2001). European Union Strategy for Sustainable Development. Adopted at the Gothenburg Summit, June 2001. https://eur-lex.europa.eu/legal-content/EN/TXT/?uri=celex:52001DC0264

[16] European Commission. (2019). The European Green Deal", COM(2019) 640 final

[17] European Commission (2018) A Clean Planet for all—A European strategic long-term vision for a prosperous, modern, competitive and climate neutral economy. COM (2018) 773

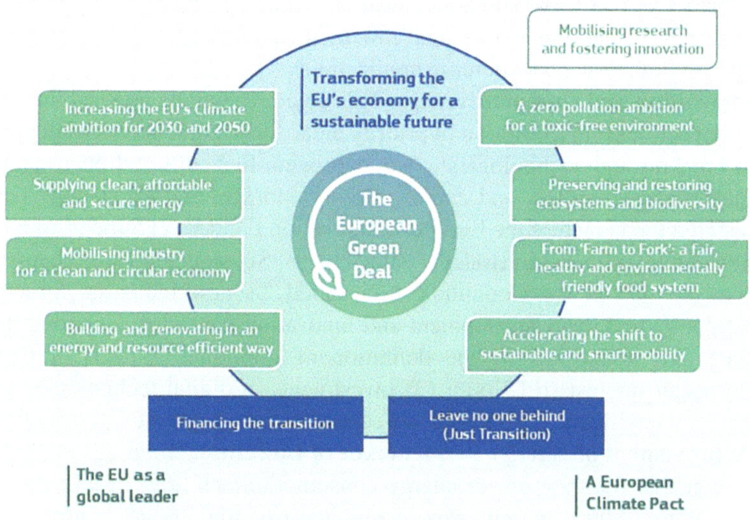

Fig. 1.2 The various elements of the European Green Deal Source: European Commission (2019) "The European Green Deal," COM(2019) 640 final, pg. 3

2. **Carbon-Content-Adjusted Price of Imports**: The EGD wants to introduce a Carbon Border Adjustment Mechanism (CBAM) for selected high-emission sectors. This mechanism should be designed to mitigate the risk of carbon leakage, which occurs when companies relocate production from the EU to countries with less stringent climate policies in order to reduce costs. By applying a carbon price to imports equivalent to that imposed on EU producers, the CBAM aims to level the playing field in international trade, discouraging offshoring of emissions.
3. **Decarbonization**: Since the production and use of energy across economic sectors account for more than 75% of the EU's greenhouse gas emissions, the EGD promotes Decarbonization patterns supporting investments in: (i) Energy Efficiency, (ii) Renewables, and (iii) a fully integration, interconnection, and digitalization of the European Energy Market to secure clean energy supply across Europe. The Decarbonization of energy-intensive industries, such as steel, chemicals, and cement[18] is considered essential.
4. **Circular Economy**: The circular economy action plan includes a policy to support the circular design of all products based on a common methodology and principles. It will prioritize reducing and reusing materials before recycling them.
5. **Not-Avoidable Waste Reduction**: Design of a robust and integrated single market for secondary raw materials and by-products through a deeper cooperation across value chains.

[18] These sectors are part of the so-called Hard-to-Abate sectors

6. **Secure Critical Raw Materials Supply:** Secure the supply of sustainable raw materials, in particular of critical raw materials necessary for clean technologies, digital, space, and defense applications.
7. **Innovation and Competitiveness:** This objective will be pursued through different actions: (1) Definition of priority areas for the first industrial deployment of breakthrough technologies: clean hydrogen, fuel cells and other alternative fuels, energy storage, and carbon capture, storage, and utilization. This aim fostered the launch of the European Innovation Fund.[19] (2) Support of cooperation and investments in strategic value chains. Six strategic value chains[20] have been defined in the Communication called "A new Industrial Strategy for Europe"[21] and their development and innovation are today supported through state aids authorized by the definition of Important Projects of Common European Interest (IPCEIs).[22] (3) Investments in digital technologies such as artificial intelligence, 5G, cloud and edge computing, and the Internet of Things.
8. **Improvement of Energy Performance of Buildings:** Buildings in the EU are responsible for 40% of our energy consumption and 36% of greenhouse gas emissions, which mainly stem from construction, usage, renovation, and demolition.[23]
9. **Sustainable and Smart Mobility:** 90% reduction in transport emissions is needed by 2050. Road, rail, aviation, and waterborne transport are expected to contribute to the reduction.
10. **"Farm to Fork":** Designing a fair, healthy, and environmentally friendly food system. This objective is expected to be achieved through: (i) a significant reduction in the use and risk of chemical pesticides, as well as the use of fertilizers and antibiotics; (ii) circular economy strategies; (iii) the promotion of sustainable food consumption.
11. **Biodiversity Preservation:** This aim is expected to be achieved by: (i) increasing the coverage of protected biodiversity-rich land and sea areas building on the Natura 2000 network[24]; (ii) increasing biodiversity in urban spaces; (iii) sustainable re- and afforestation and the restoration of degraded forests; and (iv) a sustainable "blue economy."

[19] https://cinea.ec.europa.eu/programmes/innovation-fund_en

[20] The six strategic value chains are: (1) Clean hydrogen; (2) Low-carbon industries; (3) Batteries for electromobility; (4) Renewable energy technologies; (5) Cybersecurity; (6) Cloud, edge, and quantum computing.

[21] European Commission (2020) Communication "A new Industrial Strategy For Europe", COM(2020) 102 final

[22] Both Innovation Fund and IPCEIs Framework Programme will be described in the Chapter "Funding Scouting" also a guideline to access to will be provided.

[23] https://commission.europa.eu/news/focus-energy-efficiency-buildings-2020-02-17_en

[24] Natura 2000 is an ecological network of protected areas, set up to ensure the survival of Europe's most valuable species and habitats. See: European Environment Agency (EEA) Database.

12. **Pollution Reduction:** Actions to prevent pollution from being generated as well as measures to clean and remedy it.
13. **Clean Energy Transition:** Renewable energy sources will have an essential role. Increasing offshore wind production will be essential, building on regional cooperation between Member States. The smart integration of renewables, energy efficiency, and other sustainable solutions across sectors will help to achieve decarbonization at the lowest possible cost. Today, EU funding Programmes like the Connecting Europe Facility (CEF) for Energy support this kind of projects to build new cross-border energy infrastructure in Europe or rehabilitating and upgrading the existing one.
14. **Sustainable Finance**: Mobilizing public and private investment to support the Green Deal's objectives (See Chap. 2 for a deep dive into this topic).

Focus

The Decarbonization Strategy: Charting the Path to a Sustainable Future

Industrial companies have a prominent role in achieving the Union's climate targets and climate neutrality objective by adopting carbon reduction, capture, and removal technologies and also improving European competitiveness.

Decarbonization refers to the process of reducing or eliminating the carbon dioxide (CO_2) emissions associated with a company's operations, products, and services. This is a crucial step in addressing the global challenge of climate change and transitioning toward a low-carbon economy.

To understand the significance of decarbonization, consider the case of some of the world's largest companies that have already embarked on this journey. Take, for example, Microsoft, a technology giant that has committed to becoming carbon negative by 2030. This means that Microsoft aims to remove more carbon from the atmosphere than it emits, effectively reversing its historical carbon footprint.

Another prominent example is Unilever, a consumer goods conglomerate. Unilever has set a goal of achieving net-zero emissions across its value chain by 2039, a full 11 years ahead of the Paris Agreement's target. To achieve this, Unilever is investing in renewable energy, improving energy efficiency, and collaborating with suppliers to drive down emissions across its entire supply chain. See Sect. 4.3 for a deep dive in its sustainability report.

Developing an effective decarbonization strategy requires a comprehensive approach that addresses multiple aspects of a company's operations. This typically involves:

1. Conducting a thorough carbon footprint assessment: Identifying and quantifying the main sources of emissions across the entire value chain (hot spots), from raw material extraction to product disposal (see Chap. 6 for details).

2. Setting ambitious emission reduction targets: Establishing clear, science-based goals[25] that align with global climate agreements and the company's sustainability commitments (See Sect. 6.2.1).
3. Implementing a range of decarbonization measures: Investing in renewable energy, improving energy efficiency, adopting clean and carbon capture and storage (CCS) technologies, and collaborating with suppliers and customers to drive down emissions.
4. Monitoring and reporting progress: Regularly tracking and transparently communicating the company's progress toward its decarbonization goals, ensuring accountability and driving continuous improvement.

A valuable tool for companies to identify and prioritize the most cost-effective decarbonization measures is the Marginal Abatement Cost Curve (MACC) (See Sect. 6.2.2 for details). It allows to arrange the identified abatement measures in ascending order of their marginal abatement costs (i.e., the cost per unit of emissions reduced). The MACC becomes thus an effective decision-support tool to prioritize and sequence the implementation of abatement measures based on their cost-effectiveness and emissions reduction potential.

The challenging specific objectives listed above require a structured **action plan,** i.e. a set of implementation policies. In Table 1.1, it is reported a summary of the main EU Green Deal Implementation Policies adopted by the European Commission up to date. In particular, the last column highlights the main objectives pursued by the described policy.

In order to **monitor the effectiveness of the Green deal implementation strategy,** the European Commission has also released in 2022 a visualization tool to monitor 27 key indicators for the achievement of the European Green Deal's objectives.[26] These indicators, which apply to the EU and EFTA countries are further divided into three categories: indicators that aim at reducing climate impact; indicators related to the protection of planet & health; and indicators of activities enabling a green & just transition. The list of indicators is reported in Fig. 1.3.

This chapter has outlined the European strategy for sustainability, detailing its implementation plan and the mechanisms for monitoring progress. In the European framework, each strategy serves as a foundational reference point, guiding the definition of concrete actions aimed at achieving specific sustainability targets. Effective monitoring is not merely a procedural step—it is essential to ensure that implementation remains on track, adjustments are made when necessary, and outcomes are aligned with long-term environmental and social objectives. This approach reflects the EU's commitment to accountability and continuous improvement in its sustainability agenda.

---

[25] Science-based goals refer to emission reduction targets that are aligned with the latest climate science and the global effort to limit global warming to well below 2 °C above pre-industrial levels, as outlined in the Paris Agreement

[26] https://ec.europa.eu/eurostat/cache/egd-statistics/

Table 1.1 The European action plan, the main EU Green Deal (COM(2019) 640 final) implementation policies. Source: Own elaboration

Regulation name	Reference	Year	Description	Green deal objectives[a]
A New Industrial Strategy for Europe	COM(2020) 102 final	2020	The industrial strategy supports the twin transition to a green and digital economy, make EU industry more competitive globally, and enhance Europe's open strategic autonomy	(3) Decarbonization (7) Innovation
A New Circular Economy Action Plan for a Cleaner and More Competitive Europe	COM(2020) 98 final	2020	Future-oriented agenda for achieving a cleaner and more competitive Europe in co-creation with economic actors, consumers, citizens, and civil society organizations	(4) Circular economy (5) Not-avoidable waste reduction
EU Biodiversity Strategy for 2030	COM(2020) 380 final	2020	It sets out a range of actions to protect nature and reverse the degradation of ecosystems	(11) Biodiversity
Sustainable and Smart Mobility Strategy	COM(2020) 789 final	2020	Outlines a plan to achieve a 90% reduction in transport emissions by 2050, including the promotion of electric vehicles and sustainable aviation fuels	(9) Sustainable and smart mobility
Farm to Fork Strategy	COM(2020) 381 final	2020	Aims to make food systems fair, healthy, and environmentally-friendly, including the reduction of pesticide and fertilizer use	(10) Farm to fork
Sustainable Finance Taxonomy	Regulation (EU) 2020/852	2020	Establishes a classification system to identify economic activities that are environmentally sustainable	(14) Sustainable finance
A Renovation Wave for Europe—Greening Our Buildings, Creating Jobs, Improving Lives	COM(2020) 662 final	2020	By setting minimum energy performance standards, providing financial and technical support, and addressing barriers to renovation, the EU aims to significantly improve the energy efficiency of its building stock	(8) Improvement of energy performance of buildings (1) GHG reduction
Establishing the Framework for Achieving Climate Neutrality	Regulation (EU) 2021/1119 (European Climate Law)	2021	The Regulation sets out a binding objective of climate neutrality in the Union by 2050 in pursuit of the long-term temperature goal set out in the Paris Agreement[b]	(1) GHG reduction
"Fit for 55": Delivering the EU's 2030 Climate Target on the Way to Climate Neutrality	COM(2021) 550 final	2021	In the Communication has been presented the "Fit for 55" package, i.e. a set of proposals to revise and update EU legislation in order to achieve the EU's target of reducing net greenhouse gas emissions by at least 55% by 2030	(1) GHG reduction

(continued)

Table 1.1 (continued)

EU Action Plan: "Towards Zero Pollution for Air, Water and Soil"	COM(2021) 400 final	2021	The Plan aims to ensure that by 2050, air, water, and soil pollution are reduced to non-harmful levels to health and natural ecosystems	(12) Pollution reduction
EU Adaptation Strategy	Brussels, 24.2.2021 COM(2021) 82 final	2021	The Strategy outlines a long-term vision for the EU to become a climate-resilient society, fully adapted to the unavoidable impacts of climate change by 2050. This strategy aims to reinforce the adaptive capacity of the EU and the world and minimize vulnerability to the impacts of climate change	(14) Sustainable finance
Corporate Sustainability Reporting Directive	Directive (EU) 2022/2464	2022	The CSRD is a new EU regulation that aims to improve the reporting of sustainability-related information by companies. It is an expansion and revision of the previous Non-Financial Reporting Directive (NFRD)	(14) Sustainable finance
REPowerEU Plan	COM(2022)230	2022	Its aim is to save energy, diversify energy supplies, produce clean energy rapidly reducing the dependence on Russian fossil fuels. The Commission has mobilized close to €300 billion to fund the REPowerEU Plan. The Recovery and Resilience Facility (RRF) is at the heart of this funding	(13) Clean energy transition (3) Decarbonization
The Renewable Energy Directive (RED II)	amending Directive EU/2023/2413	2023	The directive amends the previous 2030 target of at least 32% renewable energy in the EU's final energy consumption, raising it to a new target of at least 42.5% by 2030	(13) Clean energy transition
Establishing a Carbon Border Adjustment Mechanism	CBAM REGULATION (EU) 2023/956	2023	The Regulation establishes a carbon border adjustment mechanism (the "CBAM") to address greenhouse gas emissions embedded in the goods on their importation into the customs territory of the Union in order to prevent the risk of carbon leakage, thereby reducing global carbon emissions and supporting the goals of the Paris Agreement, also by creating incentives for the reduction of emissions by operators in third countries	(1) CBAM (7) Innovation and competitiveness

EU Deforestation Regulation (EUDR)	Regulation (EU) 2023/1115	2023	The EUDR aims to ensure that products sold or exported in the EU do not originate from land that was deforested or degraded after 31 December 2020	(11) Biodiversity
Net-Zero Industry Act (NZIA)	EU Regulation 2024/1735	2024	It provides a simplified regulatory framework for production capacity of key net-zero technologies like batteries, wind, solar, etc. The overarching goal is to position the EU as a global leader in the manufacturing and deployment of net-zero technologies by creating the right regulatory, funding, skills, and trade environment	(1) GHG reduction (3) Decarbonization
Clean Industrial Deal	Brussels, 26.2.2025 COM(2025) 85 final	2025	Commitment to accelerate decarbonization, reindustrialization, and innovation, all at the same time and across the entire continent, also reinforcing Europe's resilience. The focus will be mainly on two closely linked sectors. Firstly, energy-intensive industries. Secondly, the clean-tech sector. Circularity will be a priority. There are six business drivers: (1) affordable energy, (2) lead markets, (3) financing, (4) circularity and access to materials, (5) global markets and international partnerships, and (6) skills	(1) GHG reduction (7) Innovation and competitiveness (3) Decarbonization (4) Circular economy
Packaging and Packaging Waste Regulation (PPWR)	Regulation (EU) 2025/40	2025	The Regulation aims to make packaging more sustainable and to reduce the environmental impact of packaging waste by promoting a circular economy	4) Circular economy

[a]Objectives' numbering refers to the bullet point of EU Green Deal targets reported above

[b]The Paris Agreement is a legally binding international treaty on climate change. It was adopted by 196 Parties at the UN Climate Change Conference (COP21) in Paris on 12 December 2015. It entered into force on 4 November 2016

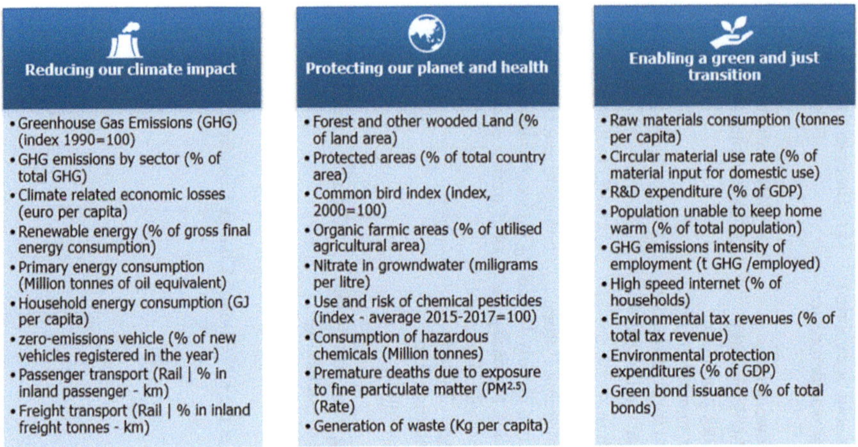

Fig. 1.3 Key indicators for the monitoring of the achievement of the European Green Deal's objectives Source: Own elaboration of the basis of https://ec.europa.eu/eurostat/cache/egd-statistics/

References

Chen, Y. C., Hung, M., & e Wang H. (2018). The effect of mandatory CSR disclosure on firm profitability and social externalities: Evidence from China. *Journal of Accounting and Economics, 65*(1), 169–190.

Council of the European Communities. (1973). *Declaration on the programme of action of the European Communities on the environment*. Publications Office of the European Union. https://op.europa.eu/en/publication-detail/-/publication/f54da63e-be4d-4664-a8ea-aa678e640b51/language-en

European Commission. (2019). *The European Green Deal, COM(2019) 640 final*.

European Council. (2001). *European Union strategy for sustainable development*. Adopted at the Gothenburg Summit, June 2001. https://eur-lex.europa.eu/legal-content/EN/TXT/?uri=celex:52001DC0264

Fortune Business Insights. (2024). *Strategic consulting market size, share & industry analysis, by service type, enterprise type, industry, and regional forecast, 2025–2032*.

Giese, G., Lee, L. E., Melas, D., Nagy, Z. E., & Nishikawa, L. (2019). Foundation of ESG Investing: How ESG affects equity valuation, risk and performance. *The Journal of Portfolio Management, 45*(5), 1–15.

Gregory, A., Tharyan, R., & Whittaker, J. (2014). Corporate social responsibility and firm value: Disaggregating the effects on cash flow, risk and growth. *Journal of Business Ethics, 124*(4), 633–657.

Management Consultancies Association. (2024). *MCA member survey 2023*. https://www.mca.org.uk/report/mca-member-survey-2023

Sustainable Finance and the Sustainability Reporting Frameworks

2

Sustainable finance refers to the process of taking environmental, social, and governance (ESG) considerations into account when making investment decisions in the financial sector, leading to more long-term investments in sustainable economic activities and projects.[1]

The concept of sustainable finance emerged from the increasing global awareness of climate change and social challenges during the late twentieth and early twenty-first centuries. Its development accelerated significantly after two landmark events: the 2015 Paris Agreement on climate change and the adoption of the United Nations 2030 Agenda for Sustainable Development.

In response, the European Commission introduced the **Action Plan on Financing Sustainable Growth**[2] in March 2018.

The Action Plan outlines a comprehensive set of measures designed to:

- Reorienting capital flows toward sustainable investments for sustainable and inclusive growth.
- Integrate sustainability into risk management and investment decision-making processes.
- Enhance transparency and long-termism in financial and economic activity.

Regarding the objective of **Reorienting capital flows toward sustainable investments for sustainable and inclusive growth**, the EU developed legislation and instruments like, among others, the EU Taxonomy Regulation (EU 2020/852),

[1] European Commission, "What is Sustainable Finance" https://finance.ec.europa.eu/sustainable-finance/overview-sustainable-finance_en

[2] European Commission. (2018). Action plan: Financing sustainable growth (COM/2018/97 final). https://eur-lex.europa.eu/legal-content/EN/TXT/?uri=CELEX%3A52018DC0097

M. Gussoni, *Integrating Strategic Sustainability Practices in Business*, CSR, Sustainability, Ethics & Governance,
https://doi.org/10.1007/978-3-032-09617-3_2

described in the Sect. 2.1.1 of the present book, which established an EU classification system for sustainable activities and the Sustainable Finance Disclosure Directive (SFDR) (EU 2019/2088) (See Sect. 2.1.2), to foster investment in sustainable projects and increasing transparency on sustainability in the financial sector.

Regarding the objective of **Integrating sustainability into risk management and investment decision-making processes,** the EU developed legislation and instruments like the Corporate Sustainability Reporting Directive (CSRD) and the European Sustainability Reporting Standards (ESRS) which requires companies to disclose detailed information on environmental, social, and governance (ESG) impacts and risks, making sustainability data clear and comparable across the EU. Reliable ESG data helps investors to make informed decisions channeling capital toward greener businesses (See Sects. 2.1.3 and 2.2).

Finally, to **foster sustainable corporate governance, enhance transparency and attenuate short-termism in capital markets,** the EU commission has adopted the Corporate Sustainability Due Diligence Directive (CSDDD) described in Sect. 2.1.4. In the CSDDD the Climate Transition Plan Requirement compels companies to integrate long-term sustainability objectives into corporate governance and decision-making.

The professional services industry has played a pivotal role in the rise of sustainable finance. Firms have developed specialized expertise in areas such as ESG risks assessment, sustainable investment strategies, and sustainability-linked financial products. They have also been instrumental in shaping the regulatory and reporting frameworks that underpin the sustainable finance ecosystem.

Sustainability reporting frameworks offer, in fact, standardized methodologies that enable organizations to communicate their environmental, social, and governance (ESG) performance transparently and consistently. These disclosures are crucial for investors and financial institutions, as they provide the necessary insights to evaluate sustainability-related risks and opportunities associated with their investments. Sects. 2.2 and 2.3 explore the European and the main international ESG and sustainability reporting frameworks highlighting approaches and guidelines that support comprehensive and credible sustainability reporting.

The reporting Frameworks described in Sect. 2.3, in particular, are the International Financial Reporting Standards for Sustainability Disclosures (IFRS—S1 and IFRS—S2) described in Sect. 2.3.1; the Global Reporting Initiative (GRI) described in Sect. 2.3.2. Finally, Sect. 2.3.3 examines regulatory developments in the United States, including the SEC's Climate Disclosure Rule and the State of California's sustainability-related reporting regulations, which reflect a growing emphasis on mandatory climate and ESG disclosures in North American markets.

2.1 The European Normative Framework for Sustainable Finance

In December 2016, the Commission mandated a High-Level Expert Group to develop an overarching and comprehensive Union strategy on sustainable finance. The report of the High-Level Expert Group published on 31 January 2018 calls for the creation of a technically robust classification system at Union level to establish clarity on which activities qualify as "green" or "sustainable" in order to foster the shift of capital flows toward more sustainable activities and avoid "greenwashing," i.e. the practice of gaining an unfair competitive advantage by marketing a financial product or an investment as environmentally friendly, when in fact basic environmental standards have not been met.

The EU Taxonomy regulation 852/2020[3] has been adopted with this aim and, thus, it plays a pivotal role in the current European Normative Framework for Sustainable Finance.

The EU Taxonomy Regulation in a nutshell:

1. It defines a classification of **environmentally sustainable economic activities** at Union level (EU Taxonomy, Art. 3): Establishing criteria for environmentally sustainable economic activities may encourage economic operators to publish and disclose information on their websites regarding the environmentally sustainable economic activities they carry out. That information will make it easier for those economic operators to raise funding for their environmentally sustainable activities.
2. It supplements, in Art. 5,6 and 7, the **rules on sustainability-related disclosures** laid down in Regulation (EU) 2019/2088 (the Sustainable Finance Reporting Directive (SFRD) on sustainability-related disclosures in the financial services sector)[4] .
3. It confirms the **obligation to publish non-financial information** pursuant to Article 19a or Article 29a of Directive 2013/34/EU (Non-Financial Reporting Disclosure Directive—NFRD)[5] on the annual financial statements, consolidated financial statements and related reports of certain types of undertakings including information on how, and to what extent the undertaking's activities are associated with economic activities that qualify as environmentally sustainable (EU Taxonomy, Art. 8). NFRD has been updated by the recent adoption of the Corporate Sustainability Reporting Directive (CSRD).[6]

---

[3] European Parliament and Council of the European Union. (2020). Regulation (EU) 2020/852 on the establishment of a framework to facilitate sustainable investment, and amending Regulation (EU) 2019/2088. Official Journal of the European Union, L 198, 13–43.

[4] European Parliament and Council of the European Union. (2019). Regulation (EU) 2019/2088 on sustainability-related disclosures in the financial services sector. Official Journal of the European Union, L 317, 1–16.

[5] European Parliament and Council of the European Union. (2014). Directive 2014/95/EU of the European Parliament and of the Council of 22 October 2014 amending Directive 2013/34/EU as regards disclosure of non-financial and diversity information by certain large undertakings and groups. Official Journal of the European Union, L 330, 1–9.

[6] European Parliament and Council of the European Union. (2022). Directive (EU) 2022/2464 of the European Parliament and of the Council of 14 December 2022 amending Regulation (EU) No 537/2014, Directive 2004/109/EC, Directive 2006/43/EC and Directive 2013/34/EU, as regards corporate sustainability reporting. Official Journal of the European Union, L 322, 15–80.

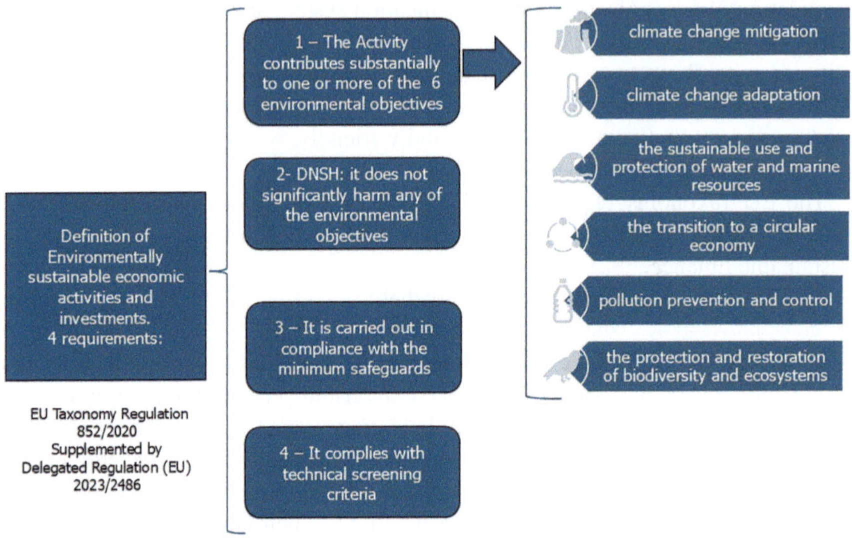

Fig. 2.1 The normative framework behind the obligation of disclosure for environmentally sustainable economic activities (Eu Taxonomy Regulation, Art 3). Source: Author elaboration

Relative to the first point, the Sustainable Economic Activities are defined as those economic activities which meet four overarching conditions (See Fig. 2.1):

1. Contribute substantially to one or more of the environmental objectives
2. Do not significantly harm any of the environmental objectives
3. Are carried out in compliance with the minimum safeguards
4. Are compliant with technical screening criteria that have been established by the European Commission

The criteria for disclosing information on environmentally sustainable economic activities under the EU Taxonomy are set out in Commission Delegated Regulation (EU) 2021/2178, which was later amended by Delegated Regulation (EU) 2023/2486. These regulations supplement Regulation (EU) 2020/852 and establish the content, format, and methodology for sustainability disclosures required under Article 8 of the Taxonomy Regulation.

The framework distinguishes between non-financial undertakings (e.g., industrial and commercial companies) and financial undertakings, which include credit institutions, asset managers, investment firms, and insurance/reinsurance companies:

- Non-financial undertakings must report the proportion of their turnover, capital expenditure (CapEx), and operating expenditure (OpEx) associated with Taxonomy-aligned activities. These indicators reflect the extent to which their core business operations contribute to environmental objectives.
- Financial undertakings, on the other hand, are subject to tailored key performance indicators (KPIs) that reflect the nature of their activities, e.g. Credit

institutions report the Green Asset Ratio (GAR), showing the share of Taxonomy-aligned exposures in their total assets, Asset managers disclose the proportion of investments in Taxonomy-aligned activities across their portfolios, Investment firms report KPIs based on both proprietary trading and client services, including revenue from Taxonomy-aligned investment activities, and Insurance and reinsurance undertakings report on both their investment portfolios and non-life underwriting activities aligned with climate adaptation goals.

The 2023 amendment (Regulation 2023/2486) introduced simplifications to reduce administrative burdens while maintaining the integrity of the framework. Together, these regulations ensure that sustainability disclosures are both sector-specific and standardized, enabling comparability and transparency across the EU's sustainable finance landscape (See next Chapter for Details).

2.1.1 The EU Taxonomy for Sustainable Activities

The Taxonomy Regulation was published in the Official Journal of the European Union on the 22nd of June 2020 and entered into force on the 12th of July 2020. The Taxonomy Regulation provides a common language and set of criteria for investors, companies, and policymakers to identify and assess the sustainability of economic activities, with the goal of directing investments toward more sustainable outcomes.

Article 9 of the Taxonomy Regulation sets out six climate and environmental objectives listed in Fig. 2.2.

1. **Climate change mitigation**: This objective aims to support economic activities that contribute substantially to the stabilization of greenhouse gas (GHG) emissions concentration in the atmosphere at a level consistent with the long-term temperature goal of the Paris Agreement.
2. **Climate change adaptation**: This objective focuses on supporting economic activities that aim to adapt to the current and future impacts of climate change, including through the prevention and reduction of change vulnerabilities and risks.
3. **Sustainable use and protection of water and marine resources**: The objective promotes economic activities that contribute to the sustainable use and protection of water and marine resources ensuring that water bodies (rivers, lakes,

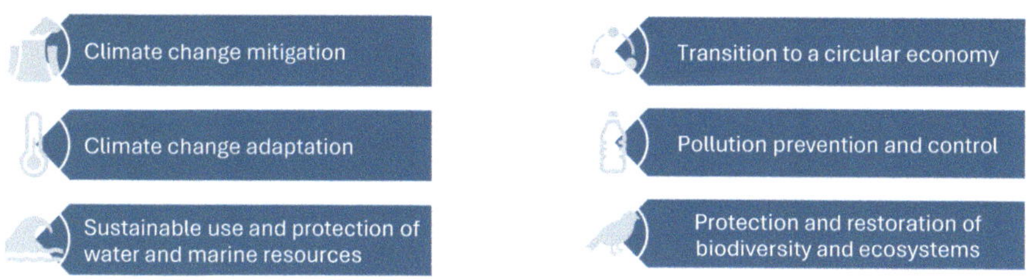

Fig. 2.2 European climate and environmental objectives. Source: Author Elaboration

groundwater) and marine ecosystems are used in a way that meets human needs without compromising their ecological integrity. This includes protecting biodiversity, maintaining water quality, and preventing overexploitation.

4. **Transition to a circular economy, waste prevention, and recycling**: This objective supports economic activities that contribute to the transition to a circular economy, including through increased recycling and waste prevention. The aim is to move away from the traditional linear model of "take-make-dispose" toward a circular model where resources are kept in use for as long as possible.
5. **Pollution prevention and control**: This objective aims to encourage economic activities that contribute to the prevention and reduction of environmental pollution. The aim is to reduce harmful emissions and contaminants across air, water, and soil to protect human health, ecosystems, and biodiversity.
6. **Protection and restoration of biodiversity and ecosystems**: This objective promotes economic activities that contribute to the protection, conservation, and restoration of biodiversity and healthy ecosystems ensuring sustainable use of resources.

To reach these objectives, the European Union has taken several actions (See Table 2.1).

Table 2.1 European actions and policies to reach the six climate and environmental objectives

Climate and environmental objectives	European actions
Climate change mitigation	• Setting of binding emissions reduction targets. • Promoting renewable energy expansion. • Promoting energy efficiency improvements. • Definition of the emissions trading system (EU ETS). • Introduction of sustainable finance regulations.
Climate change adaptation	• Adoption of the EU adaptation Strategy[a]. • Deployment of the climate-ADAPT platform[a] and the Copernicus climate change service to assess climate change vulnerabilities and risks. • Promoting of nature-based solutions.
Sustainable use and protection of water and marine resources	• Water framework Directive[c]. • Marine strategy framework Directive[b]. • Cooperation and international engagement[c].
Transition to a circular economy, waste prevention, and recycling	• Circular economy action Plan[d]. • Waste framework Directive[e]. • Packaging and packaging waste Regulation[f]. • Extended producer responsibility (EPR) schemes. • Eco-design Directive[g].

(continued)

Table 2.1 (continued)

Climate and environmental objectives	European actions
Pollution prevention and control	• Industrial emissions directive (IED)[j]; • Air quality directive;[k]
Protection and restoration of biodiversity and ecosystems	• EU biodiversity strategy for 2030[l]. • Natura 2000 network. • EU deforestation regulation (EUDR)[m].

[a]https://climate-adapt.eea.europa.eu/en

[b]European Parliament and Council of the European Union. (2008). Directive 2008/56/EC of the European Parliament and of the Council of 17 June 2008 establishing a framework for community action in the field of marine environmental policy (Marine Strategy Framework Directive). Official Journal of the European Union, L 164, 19–40

[c]The EU actively participates in international forums and agreements related to the protection of water and marine resources, such as the United Nations Convention on the Law of the Sea and regional sea conventions

[d]European Commission. (2020). A new Circular Economy Action Plan: For a cleaner and more competitive Europe (COM/2020/98 final)

[e]European Parliament and Council of the European Union. (2008). Directive 2008/98/EC of the European Parliament and of the Council of 19 November 2008 on waste and repealing certain Directives. Official Journal of the European Union, L 312, 3–30

[f]European Parliament and Council of the European Union. (2024). Regulation (EU) 2025/40 of the European Parliament and of the Council of 19 December 2024 on packaging and packaging waste, amending Regulation (EU) 2019/1020 and Directive (EU) 2019/904, and repealing Directive 94/62/EC. Official Journal of the European Union, L 22, 22 January 2025

[g]European Parliament and Council of the European Union. (2009). Directive 2009/125/EC of 21 October 2009 establishing a framework for the setting of ecodesign requirements for energy-related products. Official Journal of the European Union, L 285, 10–35

[h]European Commission. (2021). Forging a climate-resilient Europe—the new EU Strategy on Adaptation to Climate Change (COM/2021/82 final)

[i]European Parliament and Council of the European Union. (2000). Directive 2000/60/EC of the European Parliament and of the Council of 23 October 2000 establishing a framework for Community action in the field of water policy. Official Journal of the European Communities, L 327, 1–73

[j]DIRECTIVE 2010/75/EU requires industrial installations to operate under a permit system with strict environmental conditions and emission limit values. The IED requires industrial operators to use the Best Available Techniques (BAT) to prevent and control emissions. BAT reference documents (BREFs) are developed to provide guidance on the most effective techniques for different industrial sectors

[k]European Parliament and Council of the European Union. (2024). *Directive (EU) 2024/2881 of 23 October 2024 on ambient air quality and cleaner air for Europe (recast).* Official Journal of the European Union, L 2024/2881, 20 November 2024

[l]European Commission. (2020). EU Biodiversity Strategy for 2030: Bringing nature back into our lives (COM/2020/380 final)

[m]European Parliament and Council of the European Union. (2023). Regulation (EU) 2023/1115 of 31 May 2023 on the making available on the Union market and the export from the Union of certain commodities and products associated with deforestation and forest degradation and repealing Regulation (EU) No 995/2010. Official Journal of the European Union, L 150, 206–247

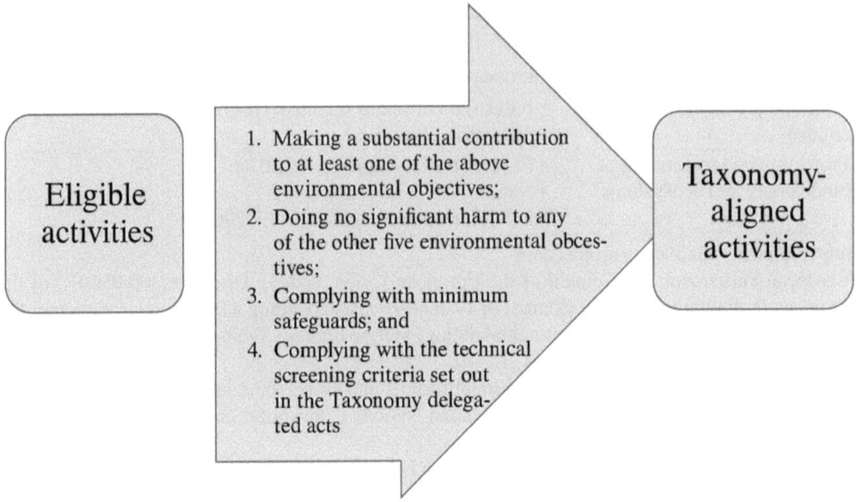

Fig. 2.3 Eligible activities and taxonomy-aligned activities

As shown in Fig. 2.1 and 2.3, according to the EU taxonomy regulation, an eligible econonomic activity, in order to qualify as **environmentally sustainable** and taxonomy aligned, must satisfy the following conditions:

1. Making a substantial contribution to at least one of the above environmental objectives;
2. Doing no significant harm to any of the other five environmental objectives;
3. Complying with minimum safeguards and;
4. Complying with the technical screening criteria set out in the Taxonomy delegated acts.

In order to satisfy condition 1, the activity shall make a substantial contribution to at least one of the environmental objectives. The EU Taxonomy Regulation defines the substantial contribution separately for the six environmental objectives considered.

Specifically:

- Article 10: Substantial contribution to climate change mitigation
- Article 11: Substantial contribution to adaptation to climate change
- Article 12: Substantial contribution to the sustainable use and protection of waters and marine resources
- Article 13: Substantial contribution to the transition toward a circular economy
- Article 14: Substantial contribution to the prevention and reduction of pollution
- Article 15: Substantial contribution to the protection and restoration of biodiversity and ecosystems (Table 2.2)

Table 2.2 Substantial contribution definition by environmental objective

Environmental objectives	Substantial contributing activities
Climate change mitigation	The activity contributes substantially to the stabilization of greenhouse gas concentrations in the atmosphere at a level which prevents dangerous anthropogenic interference with the climate system consistent with the long-term temperature goal of the Paris agreement through the avoidance or reduction of greenhouse gas emissions or the increase of greenhouse gas removals
Climate change adaptation	The activity includes adaptation solutions that either substantially reduce the risk of the adverse impact of the climate on that economic activity or substantially reduce that adverse impact, without increasing the risk of an adverse impact on people, nature, or assets; or it provides adaptation solutions that, in addition to be an enabling activity,[a] contribute substantially to preventing or reducing the risk of the adverse impact of the climate on people, nature, or assets.
Sustainable use and protection of waters and marine resources	The activity either contributes substantially to achieving the good status of bodies of water, including bodies of surface water and groundwater or to preventing the deterioration of bodies of water that already have good status, or contributes substantially to achieving the good environmental status of marine waters or to preventing the deterioration of marine waters that are already in good environmental status by: • Protecting the environment from the adverse effects of urban and industrial waste water discharges • Protecting human health from the adverse impact of any contamination of water intended for human consumption • Improving water management and efficiency • Ensuring the sustainable use of marine ecosystem services or contributing to the good environmental status of marine waters • Enabling any of the activities listed in previous points
Transition toward a circular economy	The activity: • Uses natural resources, including sustainably sourced bio-based and other raw materials, in production more efficiently • Increases the durability, reparability, upgradability, or reusability of products, in particular in designing and manufacturing activities • Increases the recyclability of products, including the recyclability of individual materials • Substantially reduces the content of hazardous substances and substitutes them with safer alternatives and ensuring traceability • Prolongs the use of products • Increases the use of secondary raw materials and their quality • Prevents or reduces waste generation • Increases preparing for the re-use and recycling of waste • Increases the development of the waste management infrastructure needed for prevention, for preparing for re-use and for recycling • Minimizes the incineration of waste and avoids the disposal of waste • Avoids and reduces litter • Enables any of the activities listed in previous points

(continued)

Table 2.2 (continued)

Environmental objectives	Substantial contributing activities
Prevention and reduction of pollution	The activity contributes substantially by: • Preventing or, where that is not practicable, reducing pollutant emissions into air, water, or land, other than greenhouse gasses • Improving levels of air, water, or soil quality • Preventing or minimizing any adverse impact on human health and the environment of the production, use, or disposal of chemicals • Cleaning up litter and other pollution • Enabling any of the activities listed in previous points
Protection and restoration of biodiversity and ecosystems	The activity contributes substantially to protecting, conserving, or restoring biodiversity or to achieving the good condition of ecosystems, or to protecting ecosystems that are already in good condition by: • Nature and biodiversity conservation • Sustainable land use and management • Sustainable agricultural practices • Sustainable forest management • Enabling any of the activities listed in previous points

[a]"enabling activity" refers to an economic activity that does not itself directly achieve an environmental objective, but facilitates or supports other activities in doing so

Moreover, the technical screening criteria described in the Annex I to the EU Taxonomy Regulation and delegated acts[7] determine, by defining some specific metrics for each sector, whether an economic activity contributes substantially to at least one of the environmental objectives (condition 1) and does no significant harm to any of the environmental objectives (condition 2).

The same categories of economic activities are subject to the same technical screening criteria for each climate objective and for this reason, the technical screening criteria follow the classification of economic activities laid down in the NACE Revision 2 classification system of economic activities.

Very often companies need support in the identification of taxonomy-eligible and taxonomy-aligned activities, and the sustainability strategy consultant provides a valuable contribution to this task. It is worth to underline, however, that also European Commission provided helpful tools to perform the job. To simplify the assessment of economic activities and their effective contribution to at least one of the environmental objectives, for example, the European Union has provided a tool: The EU Taxonomy Compass.[8] The EU Taxonomy Compass enables users to check which activities are included in the EU Taxonomy (Taxonomy-eligible activities), to which objectives they substantially contribute, and what criteria have to be met for activities to be considered taxonomy-aligned. Another tool, the Taxonomy Calculator, allows the calculation of the KPIs required for disclosure.[9]

[7] Commission Delegated Regulation 2021/2139 amended by delegated regulations 2485/2023 and 2486/2023

[8] https://ec.europa.eu/sustainable-finance-taxonomy/taxonomy-compass

[9] https://ec.europa.eu/sustainable-finance-taxonomy/

Types of Economic Activities on the basis of their Contribution to Environmental Objectives[10]

Depending on the characteristics of the activity and the technologically and economically feasible alternatives, the Taxonomy distinguishes three types of activity based on their contribution to environmental objectives. Specifically, an economic activity can:

Contribute substantially (own performance): Activities that contribute substantially according to the technical screening criteria or activities considered "green" by definition (own performance) (e.g., solar energy production).

Being an enabling activity: An economic activity is considered to contribute substantially to one or more of the environmental objectives if it directly allows other activities to make a substantial contribution to one or more of these objectives, provided that: a) it does not involve a dependence on assets that compromise the long-term environmental objectives, taking into account the economic life of such assets; and (b) has a significant positive impact on the environment, based on life cycle considerations.

For activities that contribute to the objective of mitigating climate change, **is a transition activity** an economic activity for which there are no technologically and economically viable low-carbon alternatives. It is considered to make a substantial contribution to the mitigation of climate change if it supports the transition toward a climate-neutral economy in line with a path aimed at limiting the increase in temperature to 1.5 °C compared to pre-industrial levels, including by gradually eliminating greenhouse gas emissions, in particular emissions from solid fossil fuels.

In the following it is described the condition 2, i.e. What does it mean to do no significant harm (DNSH) to any other objective the activity does not contribute to.

Do No Significant Harm Principle (DNSH)

The DNSH principle requires that an economic activity does not significantly harm any of the six environmental objectives defined in the Taxonomy Regulation and it is a crucial step in the process of determining whether an economic activity can be considered environmentally sustainable under the EU Taxonomy.

To establish whether the activity complies with this criterion, both the environmental impacts generated by the activity itself and the environmental impacts generated by the products and services provided by it are assessed, during the entire life cycle (production, use, and end of life). The evaluation takes place through the use of both quantitative and qualitative criteria.

It involves evaluating the potential negative impacts of an activity on each of the six environmental objectives and ensuring that these impacts are either negligible or can be effectively mitigated.

[10] Articles 10–16—EU Regulation 852/2020

The article 17 of the Taxonomy regulation affirms that "economic activity shall be considered to significantly harm:

- **climate change mitigation**, where that activity leads to significant greenhouse gas emissions;
- **climate change adaptation**, where that activity leads to an increased adverse impact of the current climate and the expected future climate, on the activity itself or on people, nature or assets;
- **the sustainable use and protection of water and marine resources**, where that activity is detrimental:
  - to the good status or the good ecological potential of bodies of water, including surface water and groundwater; or,
  - to the good environmental status of marine waters;
- **the circular economy**, including waste prevention and recycling, where:
  - that activity leads to significant inefficiencies in the use of materials or in the direct or indirect use of natural resources such as non-renewable energy sources, raw materials, water and land at one or more stages of the life cycle of products, including in terms of durability, reparability, upgradability, reusability or recyclability of products;
  - that activity leads to a significant increase in the generation, incineration or disposal of waste, with the exception of the incineration of non-recyclable hazardous waste; or,
  - the long-term disposal of waste may cause significant and long-term harm to the environment;
- **pollution prevention and control**, where that activity leads to a significant increase in the emissions of pollutants into air, water or land, as compared with the situation before the activity started;
- **the protection and restoration of biodiversity and ecosystems**, where that activity is:
  - significantly detrimental to the good condition and resilience of ecosystems; or,
  - detrimental to the conservation status of habitats and species, including those of Union interest."

The damage generated by economic activities to an environmental objective can be considered:

- Specific: Specific protection criteria are reported that must be respected to be considered satisfied in DNSH.
- Generic: It must refer to appendices which contain protection criteria valid for all economic activities (See Appendix A–E of the Commission's delegated regulation 2139/2021).
- Not applicable: It is considered that the specific economic activity cannot by nature causes damage to the objective considered.

Fig. 2.4 Steps for DNSH assessment. Source: Author elaboration

Given these premises, here below are the 6 key steps in performing a DNSH assessment (Fig. 2.4):

The technical criteria to perform DNSH are provided by EU Commission in Commission Delegated Regulation 2021/2139 ("Climate Delegated Act") amended by delegated regulations 2485/2023 and 2486/2023. The Taxonomy calculator is a simplified tool to both perform the Taxonomy KPIs disclosure and to do no significant harm assessment.

Minimum Safeguard Criteria

The Taxonomy Regulation sets out criteria for determining whether an economic activity can be considered environmentally sustainable. In addition to meeting the requirements outlined in Article 3, the activity must also comply with the minimum safeguards established in Article 18.

This approach aligns with the principles enshrined in the European Pillar of Social Rights,[11] which support sustainable and inclusive economic growth. It also takes into account relevant international standards for minimum human and labor rights.

The purpose of these requirements is to ensure that economic activities classified as environmentally sustainable adhere to social and labor protections, in line with the broader goals of promoting a just and equitable transition to a sustainable economy.

Minimum safeguards are to be understood as due diligence and remedy procedures implemented by a company that is carrying out an economic activity in order to ensure alignment with the Organisation for Economic Cooperation and Development Guidelines for Multinational Enterprises (OECD MNEs)[12] and the UN Guiding Principles on Business and Human Rights (UNGP).[13] The last document, in particular, provides a global standard for preventing and addressing the risk of adverse human rights impacts linked to business activities and it includes the principles and rights set out in eight of the ten fundamental conventions identified in the International Labour Organization (ILO) Declaration of the on Fundamental Principles and Rights at Work and the International Bill of Human Rights.[14]

In order to assess the non-compliance with the minimum safeguard criteria, the Platform on Sustainable Finance[15] outlines two main approaches:

1. **Lack of Human Rights Due Diligence (HRDD)**
   The company has not implemented adequate HRDD processes as required by the UN Guiding Principles and OECD Guidelines for Multinational Enterprises. A HRDD process includes:
   - Declaration, through a public document or internal policy, the commitment to responsible business conduct
   - Risk identification and assessment
   - Measures to prevent and mitigate adverse impacts
   - Monitoring and tracking effectiveness
   - Transparent communication
   - Remediation mechanisms
2. **Evidence of Human Rights Violations**
   There are credible signs that the company's failure to apply HRDD has led to human rights abuses. Reliable, impartial sources should confirm this, such as:
   - Court rulings showing the company was found liable for breaches of labor or human rights laws.

[11] European Commission. (2017). European Pillar of Social Rights. Publications Office of the European Union. https://op.europa.eu/webpub/empl/european-pillar-of-social-rights/en/

[12] https://mneguidelines.oecd.org/mneguidelines/

[13] https://www.ungpreporting.org/

[14] https://eur-lex.europa.eu/legal-content/EN/TXT/PDF/?uri=CELEX:52023XC0616(01)

[15] Platform on Sustainable Finance. (2022). Final report on minimum safeguards. European Commission.https://finance.ec.europa.eu/system/files/2022-10/221011-sustainable-finance-platform-finance-report-minimum-safeguards_en.pdf

- Stakeholder engagement failures even if it is a key requirement of the UN Guiding Principles on Business and Human Rights (UNGPs). In this case, the company fails to participate in meaningful dialogue with affected parties, especially in the context of human rights concerns.

The EU Taxonomy Calculator and the Taxonomy Disclosure

Companies that fall under the scope of the Corporate Sustainability Reporting Directive (CSRD) have to report in their annual reports to what extent their activities are covered by the EU Taxonomy (Taxonomy-eligibility) and comply with the criteria set in the Taxonomy delegated acts (Taxonomy-alignment). Other companies that do not fall under the scope of CSRD can decide to disclose this information on a voluntary basis to get access to sustainable financing or for other business-related reasons.

The reporting requirements for Taxonomy Disclosure are laid down in the Disclosures Delegated Act (Delegated Regulation (EU) 2021/2178[16] ANNEX I–ANNEX XI amended by Delegated Regulation (EU) 2023/2486[17]). The Regulations provide detailed reporting templates and instructions on how to calculate and present the taxonomy-related key performance indicators (KPIs). Taxonomy Regulation, in fact, requires companies to disclose the proportion of taxonomy-eligible and taxonomy-aligned economic activities providing specific KPIs.

According to the EU Taxonomy Regulation and the associated delegated regulations, the following types of undertakings are considered for the taxonomy disclosure requirements:

- **Non-Financial Undertakings:**
  Non-financial undertakings are companies that are not primarily engaged in financial services. They are required to disclose taxonomy-related information if they are subject to the Non-Financial Reporting Directive (NFRD) or the Corporate Sustainability Reporting Directive (CSRD).
- **Financial Undertakings:**
  Credit institutions, investment firms, insurance and reinsurance undertakings, and certain asset managers; these financial institutions are required to disclose the proportion of taxonomy-eligible and taxonomy-aligned economic activities in their total assets and the proportion of taxonomy-aligned economic activities in their lending, investment, and insurance activities.

---

[16] Supplementing Regulation (EU) 2020/852 of the European Parliament and of the Council by specifying the content and presentation of information to be disclosed by undertakings subject to Articles 19a or 29a of Directive 2013/34/EU concerning environmentally sustainable economic activities, and specifying the methodology to comply with that disclosure obligation

[17] Establishing the technical screening criteria for determining the conditions under which an economic activity qualifies as contributing substantially to the sustainable use and protection of water and marine resources, to the transition to a circular economy, to pollution prevention and control, or to the protection and restoration of biodiversity and ecosystems and for determining whether that economic activity causes no significant harm to any of the other environmental objectives and amending Commission Delegated Regulation (EU) 2021/2178 as regards specific public disclosures for those economic activities

Non-Financial Undertakings have to disclose qualitative information and KPIs related to the proportion to total turnover, capital expenditure (CapEx), and operational expenditure (OpEx) related to of environmental sustainable economic activities (Taxonomy-aligned activities) by environmental objective, expressed both in percentage terms (%) and monetary values (€).

$$KPI\,CAPEX = \frac{CAPEX\ of\ Taxonomy - aligned\ activities}{Total\,CAPEX}$$

$$KPI\,OPEX = \frac{OPEX\ of\ Taxonomy - aligned\ activities}{Total\,OPEX}$$

$$KPI\,Turnover = \frac{Turnover\ from\,Taxonomy - aligned\ activities}{Total\,Turnover}$$

These disclosures must follow the templates provided in Annex I and II of Delegated Regulation (EU) 2021/2178, as amended by Delegated Regulation (EU) 2023/2486.

Financial Undertakings have to disclose qualitative information and KPIs regarding: (i) exposures in eligible and non-eligible activities; (ii) exposures in public administrations, central banks, and supranational issuers; (iii) Derivatives; (iv) Exposures in businesses not subject to NFRD; (v) share of trading portfolio (for credit institutions); (vi) share of interbank loans on demand (for credit institutions); (vii) share of eligible and non-eligible non-life insurance activities (Economic activity 10.1 in Annex II of EU Delegated Regulation 2021/2139 "Climate Delegated Act").

In particular, the KPIs required for Financial undertakings, since January 2024, are the following (references to Delegated Reg. 2021/2178):

- Article 3 and ANNEX III describe the information to be disclosed by managers of financial assets: **Green Investment Ratio (GIR)**:

$$GIR = \frac{Taxonomy - aligned\ investments}{Total\ Investments}$$

Where:

Taxonomy-Aligned Investments (numerator) are computed as the weighted average of the value of investments in taxonomy-aligned economic activities of investee companies. A breakdown of numerator of the KPI is required.[18]

[18] The breakdown of the numerator is required distinguishing by: Taxonomy-aligned exposures to financial and non-financial undertakings subject to Articles 19a and 29a of Directive 2013/34/EU; taxonomy-aligned exposures to other counterparties.

Total Investments (denominator) are given by the Total Asset Under Management (AuM). A breakdown of denominator of the KPI is also required.[19]

- Article 4 and ANNEX V describe the information to be disclosed by Credit institutions[20]: **Green Asset Ratio (GAR)** for on-balance sheet assets[21] and KPI for off-balance sheet assets.[22] The formula to compute on-balance sheet GAR is:

$$GAR = \frac{Taxonomy - aligned\ assets}{Total\ Assets}$$

Where:

The numerator shall cover the loans and advances, debt securities, equities and repossessed collaterals, financing taxonomy-aligned economic activities based on turnover KPI and CapEx KPI of underlying assets.

The denominator shall cover the total loans and advances, total debt securities, total equities and total repossessed collaterals, and all other covered on-balance sheet assets.

GAR applying to exposures to non-financial undertakings has to be disclosed, for each environmental objective, distinguishing by:

- GAR for lending activities to equity holdings of non-financial undertakings for loans and advances (GAR L&A);
- GAR for debt securities to non-financial undertakings ("GAR DS");
- Green asset ratio for equity holdings of credit institutions in non-financial undertakings ("GAR EH");
- GAR on total financing in non-financial undertakings (lending plus equity holdings).

[19] The breakdown of the denominator is required distinguishing by: Derivatives; exposures to EU financial and non-financial undertakings not subject to Articles 19a and 29a of Directive 2013/34/EU; exposures to financial and non-financial undertakings from non-EU countries not subject to Articles 19a and 29a of Directive 2013/34/EU; exposures to financial and non-financial undertakings subject to Articles 19a and 29a of Directive 2013/34/EU; exposures to other counterparties; economic activities that are not taxonomy-eligible; taxonomy-eligible economic activities, but not taxonomy-aligned.

[20] Credit institutions shall disclose relevant KPIs on the basis of the scope of their prudential consolidation determined in accordance with Regulation (EU) No 575/2013, Title II, Chap. 2, Section 2.

[21] The on-balance sheet ratio considers only the assets and liabilities that are recorded on the organization's balance sheet.

[22] The off-balance sheet ratio considers the assets and liabilities that are not recorded on the organization's balance sheet. Examples of off-balance sheet items include operating leases, contingent liabilities, and certain types of financial instruments like derivatives.

Credit Institutions have also to disclose KPIs for off-balance sheet exposures. A complementary ratio on the level of association with taxonomy-aligned economic activities of off-balance sheet exposures that credit institutions manage and channel or contribute to channel capital flows toward economic activities whose environmental sustainability can be assessed in accordance with Regulation (EU) 2020/852. These KPIs are[23]:

- Green ratio for financial guarantees to financial and non-financial undertakings (FinGuar KPI);
- Green ratio for assets under management (AuM KPI);
- KPIs on services other than lending—Fees and Commissions (F&C KPI);
- Other disclosures in the GAR: GAR for the trading portfolio.

Article 5 and Annex VII describe the information to be disclosed by **investment firms,** i.e. KPIs on own-account investment services and activities and KPIs for management on behalf of third parties.

Investment firms dealing on their own account shall disclose the following KPIs:

KPIs related to their assets as: (i) proportion of assets associated with taxonomy-eligible economic activities within total assets, (ii) proportion of assets associated with taxonomy-aligned economic activities within assets associated with taxonomy-eligible economic activities, and (iii) proportion of assets associated with taxonomy-aligned economic activities within total assets (GAR).

Investment firms not dealing on their own account shall disclose KPIs related to revenue, including fees, commissions, and other monetary benefits as: (i) proportion of revenue from services and activities associated with taxonomy-eligible economic activities within total revenue from investment services and activities; (ii) proportion of revenue from investment services and activities associated with taxonomy-aligned economic activities within revenue from investment services and activities associated with taxonomy-eligible economic activities; and (iii) proportion of revenue from investment services and activities associated with taxonomy-aligned economic activities within total revenue from investment services and activities (GAR).

Article 6 outlines the disclosure requirements for **insurance and reinsurance undertakings**, dividing them into two categories:

- KPIs related to investments, and
- KPIs related to non-life underwriting activities.

For investment-related KPIs, insurance and reinsurance companies must report the weighted average share of investments that support or are linked to taxonomy-aligned economic activities.The term *investments* encompasses both direct and indirect holdings, including:

[23] See Annex V and Annex VI of Delegated Regulation (EU) 2021/2178 for more details

- Collective investment undertakings and participations;
- Loans and mortgages;
- Property, plant, and equipment;
- Intangible assets, where applicable.

The weighted average should reflect the proportion of taxonomy-aligned activities of the investee companies.

For non-life underwriting KPIs, insurance and reinsurance undertakings (excluding life insurers) must calculate and disclose the gross premiums written or reinsurance revenue associated with taxonomy-aligned insurance or reinsurance activities.

This KPI must be expressed as a percentage of one of the following, depending on the context:

- Total non-life insurance gross premiums written;
- Total reinsurance gross premiums written;
- Total non-life insurance revenue;
- Total reinsurance revenue.

In order to support organizations in evaluating financial metrics aligned to the Taxonomy, the EU Commission has provided an on-line tool, the **Taxonomy Calculator**.

To date, however, the tool considers only the technical criteria, set out in Annex I of Delegated Regulation 2021/2139, or the criteria for the substantial contribution to the mitigation objective and the financial metrics expected for non-financial companies, i.e. KPIs for turnover, capex, and opex.

Taxonomy Statistics

Capital investments into taxonomy-aligned activities have increased in 2024 compared to the previous year. In 2023, around 600 European companies reported capital investments into taxonomy-aligned activities of €191bn. Thus far in 2024, companies have already reported €250bn, signaling significant growth.[24]

According to Goldman Sachs and Bloomberg studies,[25] in fact, by the end of October 2024, a total of 2180 companies had disclosed their taxonomy-alignment data, representing €12.9 trillion in assets (excluding those of the financial sector). In the second year of reporting, these companies identified €250 billion in capital expenditure as taxonomy-aligned—a 34% increase

[24] EU Commission website statistics

[25] Goldman Sachs, 2025. "Equity Research," 30 January 2025

compared to 2022.[26] Over the first 2 years of reporting for the financial years 2022 and 2023, taxonomy-aligned turnover rose by 25%. The sectors with the highest shares of taxonomy-aligned turnover in 2023 were manufacturing (36%), electricity supply (33%), and construction (9%).[27]

In 2022 and 2023, companies located in Germany reported the highest taxonomy-aligned investments (€114bn), followed by France (€63bn), Spain (€60bn), and Italy (€48bn). There are evidence that the percentage of taxonomy-aligned activities is positive correlated with the market performance of companies (Cheema-Fox et al. (2023)[28]; Sautner et al. (2025)[29]).

It is worth to underline that, on July 2025, the Commission adopted a series of measures to simplify the application of the EU Taxonomy.[30]

The main simplification measures include:

- Financial and non-financial companies are exempted from assessing taxonomy eligibility and alignment for economic activities that are not financially material to their business. For non-financial companies, activities are considered irrelevant if they represent less than 10% of a company's total revenue, capital expenditure (CapEx), or operating expenditure (OpEx). Reducing this administrative burden will benefit companies, allowing them to focus on reporting and financing their core activities and how this contributes to their transition efforts.
- Furthermore, non-financial companies are exempted from assessing taxonomy alignment for all operating expenditure when it is deemed irrelevant to their business model.
- For financial companies, key performance indicators such as the Green Asset Ratio (GAR) for banks are simplified, and they are granted the option of not reporting detailed taxonomy KPIs for 2 years.

[26] EC Commission Delegated Regulation—Brussels, 4.7.2025 C(2025) 4568 final

[27] The data has been extracted from Bloomberg and Orbis databases.

[28] Cheema-Fox, A., Czasonis, M., Kontu, P., & Serafeim, G. (2023). Corporate Alignment with the EU Taxonomy for Sustainable Activities: First Evidence from Financial Accounting Data. State Street Associates.

[29] Sautner, Z., Yu, J., Zhong, R., & Zhou, X. (2025). The EU Taxonomy and the Syndicated Loan Market. Journal of Financial Services Research.

[30] EC Commission Delegated Regulation—Brussels, 4.7.2025 C(2025) 4568 final—amending Commission Delegated Regulation (EU) 2021/2178 as regards the simplification of the content and presentation of information to be disclosed concerning environmentally sustainable activities and Commission Delegated Regulations (EU) 2021/2139 and (EU) 2023/2486 as regards simplification of certain technical screening criteria for determining whether economic activities cause no significant harm to environmental objectives

- Taxonomy reporting templates are simplified, reducing the number of reported data points by 64% for non-financial companies and by 89% for financial companies.
- The criteria for "doing no significant harm" to the prevention and reduction of pollution related to the use and presence of chemicals are simplified.

The amendments are adopted in the form of a delegated act amending the delegated acts on taxonomic information and on climate and environmental regulations. The Commission published the draft of this delegated act in February 2025 as part of the Omnibus I package, allowing stakeholders to provide feedback on the proposed measures.

The simplification measures provided for in this Commission Delegated Regulation (EU) 2026/73 will apply from 1 January 2026 and will cover the 2025 financial year. However, companies have the option to apply the measures from the 2026 financial year onward, if they deem it more appropriate.

2.1.2 The Sustainable Finance Disclosure Regulation (SFDR)

Regarding financial undertakings, it is important to underline that, under Regulation (EU) 2019/2088—the Sustainable Finance Disclosure Regulation (SFDR)—financial market participants are required to disclose how they incorporate sustainability risks and factors into their investment decision-making processes. This includes providing detailed information on their sustainability policies, due diligence procedures, and the principal adverse impacts of their investment decisions on environmental and social factors. Additionally, they must report the extent to which their investments align with the EU Taxonomy. The Regulation (EU) 2019/2088 (SFDR) has been supplemented by the Level 2 framework laid out in Commission Delegated Regulation (EU) 2022/1288. Its main purpose is to provide detailed technical standards and templates for sustainability-related disclosures by financial market participants and advisors.

These regulations aims to increase transparency and consistency in the way financial market participants and financial advisors integrate sustainability risks and opportunities into their investment decisions and advisory processes. This aligns with one of the European Union's key policy goals, which is to attract private capital to support Europe's transition toward a climate-neutral economy.

The SFDR requires the following from financial market participants and financial advisers (Article 3):

1. Financial market participants must publish on their websites information about their policies on the integration of sustainability risks into their investment decision-making processes.

2. Financial advisers must publish on their websites information about their policies on the integration of sustainability risks into their investment advice or insurance advice.

The financial operators have to adopt and illustrate policies with respect to the integration of environmental, social, and governance sustainability risks (hereinafter also ESG risks—Environmental, Social, and Governance) to which investments are exposed and which may cause a negative impact on their value. They have also to provide information on how they consider the main negative effects on sustainability factors (Principal Adverse Impact, PAI, see: Art 4) that the investments themselves may determine on the environment and/or society.

The PAI disclosures are entity-level and product-level. The latter became mandatory under the Level 2 framework since December 2022 and relate to specific financial products, such as a fund, pension plan, or insurance-based investment, to inform investors how a particular product aligns with sustainability goals or considers ESG factors (Articles 8–11 of the Commission Delegated Regulation (EU) 2022/1288).

The PAI entity-level disclosures are only mandatory for Financial Market Participants (FMPs) with more than 500 employees (European Supervisory Authority, 2024).[31] The specific templates for these PAI disclosures are provided in the above mentioned Commission Delegated Regulation (EU) 2022/1288, also known as the SFDR Regulatory Technical Standards (RTS) or Level 2 Framework. These templates include a list of mandatory indicators that FMPs must disclose, covering various sustainability-related topics:

- Greenhouse Gas Emissions with reference to, among others:
  - Scope 1, 2, and 3 greenhouse gas emissions;
  - Carbon footprint;
  - GHG intensity of investee companies.
- Biodiversity and Ecosystems:
  - Investments in companies operating in or around biodiversity-sensitive areas.
- Water and Waste Management:
  - Emissions to water;
  - Hazardous waste ratio;
  - Social and Employee Matters with reference to, among others:
  - Violations of UN Global Compact principles and OECD Guidelines for Multinational Enterprises;
  - Lack of processes and compliance mechanisms to monitor compliance with UN Global Compact principles and OECD Guidelines for Multinational Enterprises;
  - Unadjusted gender pay gap;
  - Board gender diversity.

[31] European Supervisory Authorities. (2024, October 30). Final Joint ESAs 2024 report on Principal Adverse Impact disclosures under Article 18 SFDR (JC 2024 68). European Securities and Markets Authority (ESMA).

These mandatory indicators are designed to provide investors with standardized and comparable information on the principal adverse impacts of investment decisions on sustainability factors, enabling them to make more informed investment choices.

The Do No Significant Harm (DNSH) assessment applies to all sustainable investments and requires an explanation of how investments avoid harming the European environmental/social objectives.

2.1.3 The Corporate Sustainability Reporting Directive (CSRD)

The European Union has enacted legislations that mandate companies above a certain size to publicly disclose information on the risks and opportunities they perceive arising from social and environmental issues, as well as the impact of their business activities on people and the environment.

These legislations have been, first, the EU Non-Financial Reporting Directive (NFRD[32]) under which large public-interest entities (listed undertakings, banks, and insurance companies) with more than 500 employees are required to publish sustainability information. The NFRD does not mandate the use of any specific reporting framework or standard. However, if an entity chooses to rely on a particular framework—such as the Global Reporting Initiative (GRI) Standards or others[33]—it is expected to explicitly state this choice in its non-financial statement.

The Corporate Sustainability Reporting Directive (CSRD)[34] has an updated scope and reporting requirements to be phased between 2025 and 2028[35] and it has expanded, initially, to almost 50,000 the number of undertakings subject to (See Chap. 2.1.5 for the main changes introduced by the Omnibus Package). The entry into application of the reporting requirements introduced by the CSRD has been phased in the following way:

- In the first wave, large public-interest entities[36] with more than 500 employees must report for the first time in 2025, for financial year 2024.

[32] Directive 2014/95/EU of the European Parliament and of the Council of 22 October 2014 amending Directive 2013/34/EU as regards disclosure of non-financial and diversity information by certain large undertakings and groups Text with EEA relevance

[33] See Sect. 3.2.2 for details on GRI Standards

[34] Directive (EU) 2022/2464

[35] It is worth to highlight, however, that in 2025, the European Commission has adopted a new package of proposals to simplify EU rules, the so-called Omnibus packages, which is proposing changes about the timing and methods of non-financial reporting according to the indications of the CSRD (See Sect. 3.1.5).

[36] "Public-Interest Entities" are defined by Article 2 point (1) of the Accounting Directive as undertakings that are: (a) governed by the law of a Member State and whose transferable securities are admitted to trading on an EU regulated market; (b) credit institutions; (c) insurance undertakings; or (d) designated by Member States as public-interest entities.

- In the second wave, the other large undertakings must report in 2026, for financial year 2025.
- In the third wave, SMEs with securities listed in EU regulated markets must report in 2027 for financial year 2026, although they have a possibility to opt out of reporting for financial years 2026 and 2027.
- In the fourth wave, certain non-EU undertakings that have business in the territory of the Union, above certain thresholds, must report in 2029, for financial year 2028.

The rationale behind this law is to provide investors, civil society organizations, consumers, and other stakeholders with the necessary information to assess the sustainability performance of companies. By harmonizing disclosures and increasing transparency, the law aims to enable these stakeholders to make more informed decisions and hold companies accountable for their environmental and social impacts.

CSRD significantly expands the number of companies required to report sustainability information, going beyond the previous NFRD. The NFRD scope included 11,000 companies, whereas the CSRD initally applied to 42,500 companies.[37] This includes large, small, and medium public-interest entities (PIEs), listed enterprises (excluding micro-enterprises).

Unlisted European companies that meet at least two of the following three criteria are in scope:

- Total assets of at least €25 million;
- Net revenues of at least €50 million;
- Employees employed over 250.

Small and medium-sized enterprises (SMEs) that are listed on EU regulated markets are also in scope. They will follow simplified standards and have the option to opt out until 2028.

Finally, also third country companies are included if they have an European Branch generating a turnover over EUR 40 Million or if the company that owns the branch or controls the subsidiary generates more than EUR 150 Million in the EU during at least two consecutive years.

Subsidiaries of in-scope companies may be exempt from individual reporting if the parent company provides a consolidated sustainability report.

In order to report sustainability information, CSRD mandates the use of European Sustainability Reporting Standards (ESRS) (See Sect. 2.2 for details) which provide detailed indicators to be disclosed on a wide range of sustainability topics including: Environmental, Social, and Governance (ESG) Factors. This allows to improve quality, consistency, and comparability of sustainability reporting enabling stakeholders such as investors, consumers, and civil society, to make more informed

[37] Global Reporting Initiative. (2024). CSRD Essentials. Retrieved from https://www.globalreporting.org/media/nchpzct5/gri-csrd-essentials.pdf

decisions. It is also required that sustainability information must be reported in a digital format, making it more accessible and usable.

A core principle of CSRD is "Double Materiality," this means companies must report on both how their business is affected by sustainability issues (Financial Materiality) and how the business affects people and environment (Impact Materiality) (See Sect. 4.1 for details).

The concept of Impact Materiality, although not explicitly stated, was already foreseen in NFRD (Directive 2014/95/EU) which focuses on the company's impact on the outside world (*inside-out*). This impact assessment includes the computing of relevant KPIs like greenhouse gas emissions (GHG), the use of natural resources, employees working conditions, and the influence of the company on the local community. These effects must be considered and communicated, as they are relevant to a wide range of stakeholders, including customers, employees, local communities, and regulators.

The assessment of financial materiality, instead, has been introduced with the CSRD and concerns all that information that could influence investors' decisions as it focuses on the impact that the outside world has on the company (*outside-in*). This includes, for example, weather and climate risks and the potential impact that extreme weather events could have on company infrastructure, or how new environmental regulations could increase operating costs. The logic is that if something is important enough to assess a company's financial performance, then it should be reported.

A double materiality assessment is a complex and time-consuming task that requires stakeholders' engagement and in-depth analysis of their views in order to ensure greater transparency in company operations. Companies are in fact required to provide a complete and clear picture of their impacts, both internally and externally, and this transparency is essential to build and maintain trust with stakeholders.

CSRD requires companies to report on their sustainability impacts throughout their entre value chain, including upstream and downstream activities. This emphasizes the importance of supply chain management transparency. CSRD requires companies also to include in their sustainability reporting a description of their transition plans to ensure that their business model and strategy align with the transition to a sustainable economy and the objectives of limiting global warming to 1.5 °C, as well as the objective of achieving climate neutrality by 2050.

The CSRD introduces assurance requirements, meaning that reported sustainability information has to be subject to independent audits increasing reliability of the reported data.

2.1.4 The Corporate Sustainability Due Diligence Directive (CSDDD)

The Corporate Sustainability Due Diligence Directive (CSDDD) (Directive 2024/1760) has the aim to foster sustainable and responsible corporate behavior in companies' operations and across their global value chains. The directive requires companies to identify and address adverse human rights and environmental impacts of their actions and those of their business partners. The Directive also sets out an obligation for large companies to adopt and put into effect a transition plan for climate change mitigation aligned with the company's business model and strategy and with the transition to a sustainable economy.

It is worth to underline, however, that in February 2025, the Commission adopted an Omnibus package to simplify due diligence requirements (See the following Chapter for details).

The directive, before the Omnibus package, applies to both EU and non-EU companies with significant operations in the EU:

- EU Companies are subject to the directive if they account for more than 1000 employees and over €450 million in net worldwide turnover.
- Non-EU Companies are subject to the directive if they account for over €450 million in net turnover generated in the EU.
- For companies with franchise/licensing agreements in the EU, CSDDD applies if they account for royalties over €22.5 million and Net turnover over €80 million (worldwide or in the EU).

The Directive applies to companies that meet the above conditions only if those conditions are fulfilled for two consecutive financial years. The number of employees working part-time is calculated on a full-time equivalent (FTE) basis and temporary staff, and other workers employed in atypical forms of work, are included in the calculation of the number of employees as if they were workers employed directly by the company. It is important to note, however, that the regulatory framework is continuously evolving, which means the conditions outlined above may change in the coming months. Staying updated and adaptable is therefore essential for effectively navigating future requirements.

Member States have the responsibility to ensure that the subjected companies conduct risk-based human rights and environmental due diligence by carrying out the following actions:

- Integrating due diligence into their policies and risk management systems in accordance with Article 7;
- Identifying and assessing actual or potential adverse impacts in accordance with Article 8 and, where necessary, prioritizing actual and potential adverse impacts in accordance with Article 9;
- Preventing and mitigating potential adverse impacts, and bringing actual adverse impacts to an end and minimizing their extent in accordance with Articles 10 and 11;
- Providing remediation for actual adverse impacts in accordance with Article 12;
- Carrying out meaningful engagement with stakeholders in accordance with Article 13;
- Establishing and maintaining a notification mechanism and a complaints procedure in accordance with Article 14;
- Monitoring the effectiveness of their due diligence policy and measures in accordance with Article 15;
- Publicly communicating on due diligence in accordance with Article 16.

Companies must retain documentation regarding the actions carried out to fulfill their due diligence obligations for the purpose of demonstrating compliance, including supporting evidence, for at least 5 years from the moment when such documentation was produced or obtained.

Integrating Due Diligence into Company Policies and Risk Management Systems

Article 7 states that due diligence will have to be integrated by companies into their risk management policies and systems.

The policies should be developed in consultation with the company's employees and their representatives, and should contain:

- A description of the company's approach to due diligence.
- A code of conduct setting out the rules and principles to be followed not only by society as a whole, but also by its subsidiaries and direct and indirect business partners.
- A description of how due diligence procedures have been integrated into policies and how due diligence is being implemented (including measures taken to verify compliance with the Code of Conduct and its extension to business partners).

Due diligence policies will need to be updated after each significant change and, at least, every 24 months.

Identifying and Assessing Actual and Potential Adverse Impacts

According to Article 8, each company shall take appropriate measures to identify and assess the actual or potential adverse impacts caused by its own activities or those of its subsidiaries and, if linked to its chain of activities, those of its business partners.

Taking into account the relevant risk factors, companies shall take appropriate measures to:

- Map their own activities, those of their subsidiaries and, if linked to their own value chains, those of their business partners, in order to identify the areas where adverse impacts are most likely to occur and are of greatest severity.
- Carry out an in-depth assessment of their own activities and operations, those of their subsidiaries and, if linked to the value chains, those of their trading partners, in the areas identified as those where negative impacts are more likely to occur and are of greater severity. Where the information necessary for such an assessment can be obtained from business partners at different levels of the value chain, the company has to prioritize requesting such information.

Prioritization of Identified Actual and Potential Adverse Impacts

Where it is not possible to prevent, mitigate, stop, or minimize all identified adverse impacts simultaneously and comprehensively, companies must prioritize them based on the severity and likelihood. According to the defined priority, once the most severe and most likely adverse impacts have been addressed, the company can also address the less severe and less likely adverse impacts.

Preventing Potential Adverse Impacts, Bringing Actual Adverse Impacts to an End and Minimizing Their Extent or Providing Remediation for Actual Adverse Impacts

Articles 10, 11, and 12 of the directive require Member States to ensure that companies take appropriate measures:

1. To prevent, or where prevention is not possible or not immediately possible, adequately mitigate, potential adverse impacts developing and implementing a prevention action plan, with industry or multi-stakeholders engagement, with reasonable and clearly defined timelines for the implementation of appropriate measures and qualitative and quantitative indicators for measuring improvement but also asking direct business partners to make the necessary investments, improvements or upgrades (Art. 10).
2. To bring actual adverse impacts that have been identified to an end or minimize the extent of that impact (Art. 11).
3. To provide remediation where a company has caused or jointly caused an actual adverse impact (Art. 12).

Carrying Out Meaningful Engagement with Stakeholders

Article 13 requires companies to take appropriate measures to engage effectively with stakeholders.

Companies can also consult with experts who can provide credible information on actual or potential negative impacts.

Establishing and Maintaining a Notification Mechanism and A Complaints Procedure

According to Article 14, companies must allow the following entities to lodge complaints if they perceive actual or potential negative impacts on the companies' activities:

- Natural and legal persons.
- Trade unions and other workers' representatives.
- Organizations experienced in the sectors affected by the impact.

Companies will have to define complaint procedures and mechanisms (an accessible and appropriate reporting system), also ensuring anonymity and availability in response.

An example recalled by Article 30 refers to the application of EU Directive 2019/1937 (*EU Whistleblowing Directive*), concerning the protection of persons who report breaches of EU law.

Monitoring the Effectiveness of the Due Diligence Policy and Measures and Publicly Communicating on Due Diligence

Articles 15 and 16 require, respectively, companies to carry out periodic assessments of their own operations and measures, those of their subsidiaries and, where related to the value chain of the company, those of their business partners, to assess the implementation and to monitor the adequacy and effectiveness of the identification, prevention, mitigation, bringing to an end and minimization of the extent of adverse impacts by using qualitative and quantitative indicators, to be updated at least every 12 months, and report on the matters covered by this Directive by publishing on their website an annual statement.

It is worth to underline that the Omnibus proposal is amending the Corporate Sustainability Due Diligence Directive with a view to simplifying the duties and reducing regulatory burden (see the following Chapter for details) and that a set of guidelines are going to be issued by the Commission will help companies to conduct due diligence.

2.1.5 The Omnibus Package and the VSME Standard

On 26th February 2025, the European Commission has adopted a new package of proposals to simplify EU rules for sustainability reporting introduced by CSRD, the so-called Omnibus packages.

Objectives highlighted by European Commission are to boost competitiveness and simplify EU Rules (Omnibus I) and unlock additional investment capacity (Omnibus II).

Omnibus I proposals were divided into two different documents:

- Proposal for a Directive of the European Parliament and of the Council amending Directives (EU) 2022/2464 and (EU) 2024/1760 as regards the dates from which Member States are to apply certain corporate sustainability reporting and due diligence requirements.
- Proposal for a Directive of the European Parliament and of the Council amending Directives 2006/43/EC, 2013/34/EU, (EU) 2022/2464 and (EU) 2024/1760 as regards certain corporate sustainability reporting and due diligence requirements.

Omnibus I[38] postpones by 2 years the entry into application of CSRD reporting for large companies and listed SMEs to allow time for the proposed changes (it has been also called "Stop the clock directive" already adopted in April 2025 as Directive (EU) 2025/794), moreover, the new CSRD reporting requirements will only apply to large companies with over 1000 employees and either €50 million in turnover or €25 million in assets. This would reduce the number of companies in scope by around 80%.

For smaller companies outside the CSRD scope (up to 1000 employees), the Commission has developed a voluntary reporting standard (VSME) to limit the information that CSRD companies can request from their value chain and the Commission is also committed to revise the European Sustainability Reporting Standards (ESRS) to substantially reduce the number of data points, clarify provisions, and improve consistency with other legislation (See: Table. 2.3 for a summary of Changes introduced by the Omnibus I proposal) thus meeting the commitments of the Commission to reduce the administrative burden and reporting obligations by at least 25% for all business and by 35% for small and medium-sized enterprises. The explanatory memorandum indicates:

> The Commission intends to adopt a delegated act to revise the first set of ESRS. To deliver swiftly on the simplification and streamlining of the ESRS, and to provide clarity and legal certainty to undertakings, the Commission aims to adopt the necessary delegated act as soon as possible, and at the latest 6 months after the entry into force of this proposal. The revision of the delegated act will substantially reduce the number of mandatory ESRS datapoints by (i) removing those deemed least important for general purpose sustainability reporting, (ii) prioritising quantitative datapoints over narrative text and (iii) further distinguishing between mandatory and voluntary datapoints, without undermining interoperability with global reporting standards and without prejudice to the materiality assessment of each undertaking. The revision will clarify provisions that are deemed unclear. It will improve consistency with other pieces of EU legislation. It will provide clearer instructions on how to apply the materiality principle, to ensure that undertakings only report material information and to reduce the risk that assurance service providers inadvertently encourage undertakings to report information that is not necessary or dedicate excessive resources to the materiality assessment process. It will simplify the structure and presentation of the standards. It will further enhance the already very high degree of interoperability with global sustainability reporting standards. It will also make any other modifications that may be considered necessary considering the experience of the first application of ESRS.[39]

It is worth to underline that, for companies within the future CSRD scope (large companies that have more than 1000 employees) with a net turnover up to 450 million, the Omnibus proposal envisages also a voluntary Taxonomy reporting.

[38] Proposal for a Directive of the European Parliament and of the Council amending Directives (EU) 2022/2464 and (EU) 2024/1760 as regards the dates from which Member States are to apply certain corporate sustainability reporting and due diligence requirements approved, such as Directive (EU) 2025/794 "Stop the Clock"

[39] Proposal for a Directive of the European Parliament and of the Council amending Directives 2006/43/EC, 2013/34/EU, (EU) 2022/2464 and (EU) 2024/1760 as regards certain corporate sustainability reporting and due diligence requirements, Explanatory Memorandum retrieved at: https://eur-lex.europa.eu/legal-content/EN/TXT/PDF/?uri=CELEX:52025PC0081

Table 2.3 CSRD Amendment: Key elements introduced by the Omnibus I proposal

Aspect	CSRD (Directive (EU) 2022/2464)	Omnibus I Proposal (COM(2025) 80)
Scope of application	Applies to large undertakings (250+ employees, €40 M turnover, or €25 M balance sheet) and listed SMEs	Narrows CSRD scope: Applies only to companies with >1000 employees and €450 M turnover in the EU
Non-EU companies	Required to report if they generate €150 M+ turnover in the EU and have a significant EU presence	Threshold raised to €450 M EU turnover, aligning with CS3D
Reporting timeline	Phased from 2024 to 2028 depending on company size/type	Delays implementation for some companies by up to 2 years
Sector-specific standards	Commission empowered to adopt sector-specific ESRS	Removes this power; sector-specific ESRS no longer mandatory so avoiding an increase in the number of prescribed datapoints that undertakings should report
SMEs	Listed SMEs required to report under simplified standards	Voluntary reporting (VSME standard developed by EFRAG) for listed SMEs; removes mandatory requirement
Value chain reporting	Requires reporting on upstream/downstream value chain impacts	Simplifies value chain reporting; more flexibility and proportionality
Assurance requirements	Gradual move from limited to reasonable assurance	Maintains limited assurance only; no move to reasonable assurance
Administrative burden	Criticized for complexity and cost	Aims to reduce administrative burden by 25%, especially for SMEs

Source: Author elaborations on the basis of CSRD (Directive (EU) 2022/2464) and Omnibus I Proposal (COM(2025) 80)

Moreover, regarding the CSDDD, the proposal postpones by 1 year the transposition deadline (26 July 2027) and first phase of application (to 26 July 2028) to allow more time for preparation limiting the scope of due diligence to the direct business partners (Table 2.3).

Omnibus II[40] introduces amendments to the InvestEU Regulation[41] with a simplification of the procedures for accessing finance and a reduction in the documentation required for SMEs. It requires the alignment with the objectives of the Green Deal and the digital transition by prioritizing projects that contribute to environmental sustainability and technological innovation and increased transparency and monitoring with the introduction of simplified indicators for impact assessment.

[40] Proposal for a Regulation of the European Parliament and of the Council amending Regulations (EU) 2015/1017, (EU) 2021/523, (EU) 2021/695 and (EU) 2021/1153 as regards increasing the efficiency of the EU guarantee under Regulation (EU) 2021/523 and simplifying reporting requirements

[41] Regulation (EU) 2021/523 of the European Parliament and of the Council of 24 March 2021 establishing the InvestEU Programme and amending Regulation (EU) 2015/1017

One of the main impacts of the Omnibus Package is the reduction of the number of companies falling in the CSRD scope. These companies are however encouraged to communicate ESG data. For this reason, the EU Commission had given the mandate to EFRAG of developing a **Voluntary Reporting Standard for non-listed micro, small, and medium enterprises (VSME).**[42] The VSME is a reporting standard for non-listed micro, Small, and Medium Enterprises (SMEs) which do not fall within the mandatory reporting requirements of the Corporate Sustainability Reporting Directive (CSRD). The VSME aims to redefine ESG used by banks and large companies to collect data from SMEs. According to EFRAG, SMEs that adopt ESG practices and report them can, in fact, benefit from subsidized loans with discounts of up to 30 basis points.[43]

The VSME is aimed to satisfy data needs from larger undertakings and banks and is organized in two modules:

- The basic module;
- The comprehensive module.

The basic modules is composed of 11 disclosures both qualitative and quantitative, the comprehensive module adds 9 disclosures on top of the basic module. For both modules, the disclosures are spread across General Information, Environment, Social and Business conduct Metrics. No materiality Matrix is required.

In the Basic Module, 7 out of 11 Datapoints are mandatory and regard the basis of the sustainability report preparation, including whether the report is consolidated or individual, and the list of subsidiaries covered; Energy and Greenhouse Gas emissions; Water; resources use circular economy and waste management and workforce.

The Comprehensive Module requires nine additional disclosures, of environmental (2), social (3), governance (2), and cross-cutting (2) nature. The Comprehensive Module, on top of the Basic module, can satisfy the data needs of both banks and corporate customers. Even in this case, 4 datapoints out of 9 are mandatory and regard Strategy, Human Rights, and Revenues.

2.2 The European Sustainability Reporting Standards (ESRS)

The first set of European Sustainability Reporting Standards (ESRS) has been adopted as delegated act (EU Commission Delegated Regulation 2023/2772) and published in the European Commission Official Journal by the end of 2023.

[42] https://www.efrag.org/sites/default/files/sites/webpublishing/SiteAssets/VSME%20 Standard.pdf

[43] EFRAG (2024) EFRAG's Cover letter for the Cost-Benefit Analysis on the voluntary sustainability Reporting Standard for Non-listed SMEs (VSME)

The ESRS ensures the quality and the consistency of the reported information, by requiring that it is understandable, relevant, verifiable, comparable, and represented in a faithful manner providing a comprehensive guideline to be compliant with the CSRD Requirements.

ESRS are based on technical advice from EFRAG, a private association established in 2001 with the encouragement of the European Commission.[44]

Companies falling in the CSRD scope should disclose information on five reporting areas:

- Business model;
- Policies, including the due diligence processes implemented;
- The outcome of these policies;
- Risks management;
- Key performance indicators relevant to the business.

There are three categories of ESRS: (a) cross-cutting standards; (b) topical standards (Environmental, Social, and Governance standards); and (c) sector-specific standards.

Cross-cutting standards are sector-agnostic, i.e. they are applicable to all the type of undertakings regardless the sector they belong to. They are ESRS 1 "General requirements" and ESRS 2 "General disclosures."

ESRS 1 describes the objective and the structure of ESRS and the general requirements for preparing and presenting sustainability-related information. Among other, ESRS 1 requires all companies to conduct a materiality assessment that applies the principle of double materiality.

ESRS 2 is the standard that outlines general disclosure requirements for sustainability reporting under CSRD. It applies across all sustainability topics and sets the foundation for how companies should report. In other words, it establishes Disclosure Requirements, and also specifies the Minimum Disclosure Requirements (MDR) on the information that the company shall provide at a general level, across all material sustainability matters. MDR are baseline disclosures that companies must provide, regardless of the specific sustainability topic. They are clustered in MDR-A related to Actions, MDR-P related to Policies, MDR_M related to Metrics, and MDR-T related to Targets.

According to these standards, companies have to describe how they affect the environment and the social community; any actions taken, and the result of such actions in preventing or mitigating actual or potential negative impacts. It is also required to describe the plans and capacity of the company to adapt its strategy and business model in line with the transition to a sustainable economy and the financial effects on the company of material risks and opportunities related to sustainability matters.

---

[44] A detailed description of ESRS Set 1 and guidelines provided by EFRAG is available at: https://xbrl.efrag.org/e-esrs/esrs-set1-2023.html#d1e134-3-1

ESRS 1 and ESRS 2 apply to sustainability matters described in both the topical standards and sector-specific standards. The information requested by ESRS 2 is mandatory for all companies. The support tool generally used to check the completeness of information requested by ESRS2 is the EFRAG Implementation Guidance 3 List of Datapoint Excel Workbook[45]; it is thus suggested to perform a gap analysis in order to be compliant with the mandatory requirements.

Unlike ESRS 2, all other standards are subject to a materiality assessment. The materiality assessment identifies material Impacts, Risks and Opportunities (IROs) that should be reported in the company's own operations and its value chain. It is worth to underline that the term "impacts" refers to positive and negative sustainability-related impacts.

Sector-specific standards address impacts, risks, and opportunities that are likely to be material for all the companies belonging to a specific sector. By addressing sector-specific IROs, the organizations report information that are specific to that sector, and that are not already covered by the topical ESRS.

Topical ESRS (10 indicators) reflect the three dimensions of sustainable development (ESG). Each dimension is indicated by a letter and a number ESRS E_i (with $i = 1\ldots5$) refer to the environmental dimension, ESRS S_j (With $j = 1\ldots4$) refer to the social dimension, and ESRS G_1 to the Governance one.

The Disclosure Requirements in ESRS 2, in topical ESRS and in sector-specific ESRS are structured into the following reporting areas:

- Governance (GOV): Description of the governance processes, controls, and procedures used to monitor, manage IROs.
- Strategy (SBM): Description of how strategy and business model interact with IROs.
- Impact, Risk, and opportunity Management (IRO): Description of the process by which the company assess IROs Materiality and manage material issues through policies and actions.
- Metrics and Targets (MT): Description of the company's performance targets and metrics to measure progress toward meeting them.

2.2.1 Environmental Dimension Topical ESRS

The Topical ESRS E1 Standard Refers to Climate Change and it covers Disclosure Requirements related to the following sustainability matters: Climate Change mitigation, Climate Change Adaptation, and Energy Related Matters.

The disclosure related to climate change mitigation requires the company to compute GHG emissions of its industrial processes and to describe how it addresses these emissions through mitigation efforts in line with the Paris Agreement and compatible with limiting global warming to 1.5 °C.

[45] Source: https://www.efrag.org/en/projects/esrs-implementation-guidance-documents

The climate change adaptation requires the company to describe how it is addressing climate change actual and expected risks. Finally, energy related disclosure regards energy production and consumption topics.

The following figure presents all disclosure requirements under ESRS E1, accompanied by an overview of how sustainability strategy consultants can support companies in achieving compliance (Table 2.4).

Table 2.4 Disclosure requirements for ESRS E1 and the supporting strategic consulting services

Disclosure requirements E1	Requirement description	Strategic consulting services
E1–1: Transition plan for climate change mitigation	It provides a high-level explanation of how the company will adjust its strategy and business model to ensure compatibility with the transition to a sustainable economy and with the limiting of global warming to 1.5 °C in line with the Paris agreement and the objective of achieving climate neutrality by 2050. Companies are required to disclose climate-related scenario analysis to report on their climate resilience and identify climate-related risks and opportunities. Companies need to demonstrate how the insights from scenario analysis are integrated into their overall strategy and decision-making processes.	Consultants can support companies in the transition plan definition considering also sectoral pathways and the limiting of global warming to 1.5 °C. Consultants provide expert guidance on the methodologies and best practices for conducting scenario analysis. They also help integrate the insights from scenario analysis into the company's overall strategy and decision-making processes[a].
E1–2: Policies related to climate change mitigation and adaptation	Companies need to disclose their policies for mitigating and adapting to climate change.	Consultants can assist in formulating and documenting these policies, ensuring they are robust and actionable.
E1–3: Actions and resources in relation to climate change policies	Companies must disclose the actions taken and resources allocated to implement their climate change policies.	Consultants can help identify and prioritize actions, allocate resources effectively, and document these efforts for disclosure.
E1–4: Targets related to climate change mitigation and adaptation	Companies need to disclose their specific targets for mitigating and adapting to climate change. They have to assess compatibility of the transition plan with the objective of limiting global warming to 1.5 °C and provide information related to climate-related hazards	Consultants can support clients in the calculation of the GHG emissions (scopes 1, 2, and 3—See Sect. 6.2) e support in setting GHG emission reduction targets. The company shall disclose its current base year and baseline value, and from 2030 onward, update the base year for its GHG emission reduction targets after every 5-year period thereafter, if possible until 2050. Consultant can support also in Climate-Risk assessment and elaboration of mitigation measures.

(continued)

Table 2.4 (continued)

Disclosure requirements E1	Requirement description	Strategic consulting services
E1–5: Energy consumption and mix	Companies must disclose their total energy consumption and the mix of energy sources used.	Consultants can help in accurately measuring and reporting energy consumption, and in developing strategies to increase the use of renewable energy sources.
E1–6: Gross scopes 1, 2, 3 and total GHG emissions	Companies need to disclose their greenhouse gas (GHG) emissions across all scopes.	Consultants can support clients in calculating and reporting GHG emissions, and in identifying opportunities for emission reductions
E1–7: GHG removals and GHG mitigation projects financed through carbon credits	Companies must disclose any GHG removals and mitigation projects financed through carbon credits.	Consultants can help identify and implement GHG removal projects and navigate the carbon credit market (see Sect. 6.2.3)
E1–8: Internal carbon pricing	Companies need to disclose their internal carbon pricing mechanisms	Consultants can assist in developing and implementing internal carbon pricing strategies. Internal carbon pricing (ICP) is a mechanism where companies assign a monetary value to their greenhouse gas (GHG) emissions. This price is used internally to guide decision-making and incentivize emission reductions. By placing a cost on emissions, ICP encourages business units to reduce their carbon footprint. This can steer investments toward low-carbon technologies and practices.
E1–9: Anticipated financial effects from material physical and transition risks and potential climate-related opportunities	Companies should provide information related to climate-related hazards that may lead to physical climate risks and possible adaptation solutions for reducing these risks. The disclosure should also cover transition risks.	Consultant can support in Climate-Risk assessment and elaboration of mitigation measures, moreover, they can help in assessing and quantifying financial effects, integrating them into financial planning and reporting

Source: Author elaboration on the basis of EFRAG Guidelines

[a]For the scenario analysis, it may be considered the following guidance: TCFD Technical Supplement on "The Use of Scenario Analysis in Disclosure of Climate-Related Risks and Opportunities" (2017); TCFD "Guidance on Scenario Analysis for Non-Financial Companies" (2020); ISO 14091:2021 "Adaptation to climate change—Guidelines on vulnerability, impacts and risk assessment"

The Topical ESRS E2 Standard Refers to Pollution. The companies are asked to disclose how they affect pollution of air, water, and soil, in terms of material positive and negative actual or potential impacts; any actions taken, and the result of such actions, to prevent or mitigate actual or potential negative impacts, and to address risks and opportunities; the plans and capacity to adapt their strategy and business model in line with the need to prevent, control, and eliminate pollution; financial effects over the short-, medium-, and long-term of material risks and opportunities arising from pollution-related impacts and dependencies.

In the following table, as already done for ESRS E1, it will be reported all the ESRS E2 disclosure requirements and a description of how strategic consulting could support companies in this task (Table 2.5).

Table 2.5 Disclosure requirements for ESRS E2 and the supporting strategic consulting services

Disclosure requirements E2	Requirement description	Consulting services
E2–1: Policies related to pollution	The company is required to describe Policies adopted to manage its material impacts, risks, and opportunities related to pollution prevention and control	Consultants can help to develop and implement comprehensive pollution management policies. They can also assist in integrating these policies into the company's overall sustainability strategy
E2–2: Actions and resources related to pollution	This requires companies to detail the actions taken and resources allocated to prevent, control, and reduce pollution	Consulting services can aid in planning and executing actions to prevent, control, and reduce pollution. This includes identifying necessary resources, optimizing their allocation, and ensuring effective implementation.
E2–3: Targets related to pollution	Companies need to set and disclose specific targets for reducing pollution	Consultants can assist companies in setting realistic and achievable pollution reduction targets. They can also develop systems for monitoring progress and ensuring targets are met.
E2–4: Pollution of air, water, and soil	Companies must report on their emissions into air, water, and soil, including measures taken to mitigate these emissions[a]. They also must report the microplastics it generates or uses.	Strategic consultants can help accurately report emissions and develop effective mitigation strategies. This includes leveraging technology and best practices to reduce pollution

(continued)

Table 2.5 (continued)

Disclosure requirements E2	Requirement description	Consulting services
E2–5: Substances of concern and very high concern	Disclosure of substances that pose significant risks to health and the environment, including any restrictions on their use	Consulting services can provide expertise in identifying and managing risks associated with substances of concern. They can also help navigating regulatory requirements and restrictions
E2–6: Anticipated financial effects from pollution-related impacts	Companies should disclose the financial implications of pollution-related risks and opportunities over short-, medium-, and long-term horizons	Consultants can conduct detailed analyses of the financial impacts of pollution-related risks and opportunities. This helps companies understanding the economic implications and making informed decisions.

Source: Author elaboration on the basis of EFRAG Guidelines
[a]Companies have to report each pollutant listed in Annex II of Regulation (EC) No 166/2006 of the European Parliament and of the Council

The Topical ESRS E3 Standard Refers to Water and Marine Resources. Companies are required to report how they affect water and marine resources, in terms of material positive and negative actual or potential impacts; any actions taken, and the result of such actions to prevent or mitigate material negative impacts and to protect water and marine resources, also with reference to reduction of water consumption.

Companies should also report their contributes to the European Green Deal's ambitions for fresh air, clean water, healthy soil, and biodiversity, as well as to the sustainability of the blue economy and fisheries sectors (See Chapt. 1). They should also report about their capacity to adapt their business models in line with the promotion of sustainable water use based on long-term protection of available water resources; protection of aquatic ecosystems and restoration of freshwater and marine habitats. Scaling up and accelerating the effective implementation and enforcement of the existing EU water acquis, as further strengthened over the last 5 years in the context of the European Green Deal is, in fact, an essential pillar of EU's efforts toward reaching water resilience. It is worth noting that, while water is abundantly present on earth, freshwater makes up to only 2.5% of the total global water resources, and only 1% of that freshwater is available for human consumption because the vast majority is either locked in glaciers and ice and the rest is deep underground. The freshwater consumption worldwide is driven by irrigated agriculture, which accounts for roughly 70% of freshwater consumption, followed by municipal water demand. Looking at the energy sector, the amount of water withdrawal for primary energy production and power generation is roughly 10% of global freshwater (IEA Water Energy Nexus, 2016).[46]

[46]International Energy Agency. (2016). Water-Energy Nexus: World Energy Outlook Special Report. https://www.iea.org/reports/water-energy-nexus

Given these premises it becomes crucial a strong commitment toward the preservation of water and marine resources. At this regard, the ESRS E3 outlines several key disclosure requirements. Strategic consulting services can provide valuable support to companies in meeting these requirements effectively as reported in the following table (Table 2.6).

Table 2.6 Disclosure requirements for ESRS E3 and the supporting strategic consulting services

Disclosure requirements E3	Requirement description	Consulting services
E3–1: Policies related to water and marine resources	Companies must disclose their policies for managing water and marine resources. Water management includes: (i) water use; (ii) water treatment; (iii) prevention of water pollution. Companies should also demonstrate the commitment to reduce water consumption in areas at water risk in its own operations and along the value chain.	Consultants can help develop and implement comprehensive policies that align with ESRS E3–1 disclosure requirements.
E3–2: Actions and resources related to water and marine resources	Companies should provide details on actions taken and resources allocated to manage water and marine resources.	Consulting services can assist in planning and executing actions, identifying necessary resources, and optimizing their allocation.
E3–3: Targets related to water and marine resources	Companies must define and transparently disclose clear, measurable targets for the sustainable management of water and marine resources, ensuring full alignment with ESRS E3 principles and demonstrating accountability	Consultants can help to set realistic and achievable targets, and develop systems for monitoring progress.
E3–4: Water consumption	Companies should report on water consumption, including measures taken to reduce it. The disclosure must include: (i) water consumption distinguishing by area of high-water stress; (ii) water recycling; (iii) water stored. Companies should also provide information on their water intensity[a]	Strategic consultants can aid in accurately reporting water consumption, water recycling and storage and developing effective strategies to optimize water use.
E3–5: Anticipated financial effects from water and marine resources-related impacts	Companies should disclose the financial implications of water and marine resources-related risks and opportunities.	Consultants can conduct detailed analyses of the financial impacts of an inefficient water management, helping companies in understanding the economic implications and making informed strategic decisions.

Source: Author elaboration on the basis of EFRAG Guidelines
[a]Water intensity is defined as the total water consumption in the company's own operations in m^3 per million EUR net revenue (Source: ESRS E3)

The Topical ESRS E4 Standard Refers to Biodiversity and Ecosystems. The companies are required to disclose how their operations affect biodiversity and ecosystems, actions taken to mitigate negative impacts, and how their strategy and business models are aligned to respect planetary boundaries related to biosphere integrity and land-system change.[47]

The Table below reports disclosure requirements under ESRS E4, along with their descriptions and associated consulting services (Table 2.7).

Table 2.7 Disclosure requirements for ESRS E4 and the supporting strategic consulting services

Disclosure requirements E4	Requirement description	Consulting services
E4–1: Transition plan and consideration of biodiversity and ecosystems in strategy and business model	The companies should outline how biodiversity and ecosystems are integrated into their own strategy and business model.	Consulting services related to this disclosure requirement include strategic planning considering the integration of sustainability issues, and transition management consulting. The definition of a transition plan aligned with the EU biodiversity strategy for 2030, the global biodiversity framework (GBF), and the sustainable development goals (SDGs) is a first step toward a more sustainable pattern as well as the definition of a business model that embeds biodiversity issues at its core, ensuring resilience and long-term value creation.
E4–2: Policies related to biodiversity and ecosystems	Companies should describe the policies related to biodiversity and ecosystems. In particular, companies should disclose if they adopted biodiversity and ecosystem protection policy covering operational sites owned, leased, or managed in or near a biodiversity-sensitive area and policies to address deforestation.	Consulting services related to this disclosure requirement include the support to policy development, compliance consulting, and environmental policy advisory.
E4–3: Actions and resources related to biodiversity and ecosystems	Companies should detail the actions taken and resources allocated to manage biodiversity and ecosystems.	Consulting services associated to this disclosure requirement include Project management, resource allocation consulting, and implementation support.

(continued)

[47] Land-system change addresses the sustainable use and management of land to prevent degradation and loss of natural habitats

Table 2.7 (continued)

Disclosure requirements E4	Requirement description	Consulting services
E4–4: Targets related to biodiversity and ecosystems	Set and disclose targets related to biodiversity and ecosystems.	Among consulting services associated to this disclosure we can include Target setting, performance measurement, and goal tracking consulting. Some targets to be set can be, for instance, reduction in land use change, reduction in species extinction risk, improvement in ecosystem condition and function, hectares of habitat restored, number of biodiversity-positive projects implemented, etc.
E4–5: Impact metrics related to biodiversity and ecosystems change	Report on metrics related to the impact on biodiversity and ecosystems.	Among consulting services associated to this disclosure we can include Data analytics, impact assessment, and reporting services.
E4–6: Financial effects	Disclose the anticipated financial effects from biodiversity and ecosystem-related risks and opportunities.	Consultant can support clients in providing financial risk assessment, economic impact analysis, and sustainability reporting consulting.

Source: Author elaboration on the basis of EFRAG Guidelines

The Topical ESRS E5 Standard refers to Resource Use and Circular Economy. The Objective of this standard is that of specifying the Disclosure Requirements that help stakeholders in understanding how companies affects resource use, including resource efficiency, avoiding the depletion of resources and the sustainable sourcing and use of renewable resources.

Any actions taken to mitigate the negative actual or potential impacts and the plans of the companies to adapt their strategies and business models in line with circular economy principles, including minimizing waste and maintaining the value of product and resources, enhancing their efficient use in production and consumption, have to be reported. The Standard requires companies also to disclose material risks and opportunities and estimate their financial effects over the short, medium, and long term.

The Figure below reports disclosure requirements under ESRS E5, along with their descriptions and associated consulting services in support (Table 2.8).

2.2.2 Social Dimension Topical ESRS

The Topical ESRS S1 Standard Refers to Own Workforce. The objective of this Standard is to specify disclosure requirements which will enable stakeholders to understand the company's material impacts on its own workforce, as well as related material risks and opportunities.

Table 2.8 Disclosure requirements for ESRS E5 and the supporting strategic consulting services

Disclosure requirements E5	Requirement description	Associated consulting services
E5–1: Policies related to resource use and circular economy	Companies must describe policies adopted to manage their material impacts, risks, and opportunities (IROs) related to resource use and circular economy.	Consultants can support clients in the required policy development also on the basis of the identified IROs
E5–2: Actions and resources related to resource use and circular economy	Key actions taken and planned to achieve the resource use and circular economy-related policy objectives and targets have to be described.	Among strategic consulting services related to this disclosure requirement: Project management, resource allocation consulting, and implementation support. Business planning support. Definition of a strategic roadmap to set circular business models (See Chapter 7.1 for details).
E5–3: Targets to resource use and circular economy	Companies must set and disclose targets related to resource use and circular economy.	Associated consulting services: Target setting, performance measurement, monitoring and goal tracking consulting.
E5–4: Resource inflows	Companies must report on the inflows of resources into the organization. This disclosure requirement has the aim to enable an understanding of the resource use in the undertaking's own operations and its upstream value chain. The total weight of resource inflows, the percentage of biological materials, the weight or the percentage of secondary materials have to be provided.	Associated consulting services: Data analytics, supply chain management, and resource tracking consulting. Life cycle assessment (LCA) (see Sect. 6.1)
E5–5: Resource outflows	Report on the outflows of resources from the organization including wastes. The company shall describe how it contributes to circular economy and waste reduction and management strategies.	Associated consulting services: Waste management, circular economy consulting, and resource optimization services. Circular business model Design[a]. Life cycle assessment (see Sect. 6.1)
E5–6: Financial effects	Disclose the anticipated financial effects from resource use and circular economy-related risks and opportunities.	Associated consulting services: Financial risk assessment, economic impact analysis, and sustainability reporting consulting.

Source: Author elaboration on the basis of EFRAG Guidelines
[a]See Sect. 7.1 for details about circular business model design strategic consulting services

Material, actual, and potential impacts have to be assessed in relation to:

- **Working conditions**: Secure employment; working time; adequate wages; social dialogue; freedom of association, the existence of works councils and the information, consultation and participation rights of workers; collective bargaining, including the rate of the undertaking's workforce covered by collective agreements; work–life balance; and health and safety.
- **Equal treatment and opportunities:** Gender equality and equal pay for work of equal value; training and skills development; employment and inclusion of persons with disabilities; measures against violence and harassment in the workplace; and diversity.
- **Other work-related rights:** Child labor, forced labor, etc.

The Table below reports disclosure requirements under ESRS S1, along with their descriptions and the associated consulting services (Table 2.9).

The Topical ESRS S2 Standard Refers to Workers in the value chain. The objective of this standard is to specify disclosure requirements which will enable

Table 2.9 Disclosure requirements for ESRS S1 and the supporting strategic consulting services

Disclosure requirement	Requirement description	Associated consulting services
S1–1 Policies related to own workforce	Companies must disclose policies on employment, working conditions, diversity, etc.	Strategic sustainability consultants can support clients in Policy review and development and HR compliance audits
S1–2 Engagement with workers	Companies must describe processes for engaging with employees and representatives.	The consultants usually support their clients in defining the Stakeholder engagement strategy. Preparation of surveys and questionnaire and data elaboration can also facilitate the Stakeholder consultation (See Sect. 4.2 for details on stakeholders mapping and engagement).
S1–3 Remediation and grievance mechanisms	Companies should explain how possible negative impacts on own workforce are addressed and how workers can raise concerns.	Associated consulting services: Grievance mechanism design, human rights due diligence.
S1–4 Actions on material impacts	Companies should report on the identified material Impact, risks and Opportunities (IROs) and the actions taken to mitigate risks and enhance opportunities	Strategic sustainability consultants usually help in the double materiality assessment process (See Sect. 4.1 for details) and IROs identification, as well as in the definition of risk mitigation plans.
S1–5 Targets and performance	Companies have to set and disclose targets for workforce-related impacts and risks	Associated consulting services: KPI development, sustainability performance tracking

(continued)

Table 2.9 (continued)

Disclosure requirement	Requirement description	Associated consulting services
S1–6 Employee characteristics	Companies have to provide demographic and employment data.	Associated consulting services: Workforce analytics, HR data management
S1–7 Non-employee worker characteristics	Companies should disclose also data on temporary, agency, or other non-employee workers	Associated consulting services: Workforce analytics, HR data management, Contractor management systems, labor supply chain audits.
S1–8 Collective bargaining and dialogue	Companies should report on union coverage and social dialogue practices	Consultant can support clients in Labor relations collective bargaining.
S1–9 Diversity metrics	The disclosure requirement focuses on providing transparency about workforce diversity within the company's own workforce. Companies should provide data on gender, age, and other diversity indicators. In particular, the undertakings must disclose the gender distribution at top management and the age distribution amongst their employees.	Strategic consultants can help clients in defining a DEI strategy[a] and providing diversity reporting tools
S1–10 adequate wages	Companies should disclose whether or not their employees are paid an adequate wage, and if they are not all paid an adequate wage, the countries and percentage of employees concerned. It is necessary to disclose wage levels and adequacy assessments.	Consultants can provide wage analyses and compensation benchmarking statistics.
S1–11 social protection	Companies should disclose whether their employees are covered by social protection against loss of income due to major life events, and, if not, the countries where this is not the case. It is necessary the companies report on access of their own workforce to benefits like healthcare, pensions, etc.	Associated consulting services: Benefits program evaluation, social protection gap analysis aiming to assess and improve coverage. Collection of data on existing social protection schemes, coverage rates, benefit adequacy, and accessibility.
S1–12 persons with disabilities	The undertakings should disclose the percentage of their own employees with disabilities and disclose inclusion practices and data on disabled workers	Associated consulting services: Accessibility audits systematically evaluating how well a physical space, digital product, or service meets the needs of people with disabilities. Support to the development of an inclusive hiring strategy creating a recruitment process that actively promotes inclusion across all stages—From job design to onboarding.

(continued)

Table 2.9 (continued)

Disclosure requirement	Requirement description	Associated consulting services
S1–13 training and skills development	The undertakings should disclose the extent to which training and skills development is provided to its employees. Statistics on training should be broken down by gender.	Consultants support clients in reporting on training hours, programs, and outcomes and in developing learning and development strategy.
S1–14 health and safety	Companies should disclose information on the extent to which their own workforce is covered by their health and safety management system and the number of incidents associated with work-related injuries, ill health and fatalities of their own workforce.	Associated consulting services: Occupational health audits, safety program design.
S1–15 work–life balance	Companies should disclose practices supporting flexible work and Well-being	Associated consulting services: Work–life balance assessments, flexible work policy design.
S1–16 compensation metrics	Companies should disclose the percentage gap in pay between its female and male employees and the ratio between the remuneration of its highest paid individual and the median remuneration for its employees.	Consultants can support clients in reporting on pay gaps and total compensation. They can also perform Pay equity analyses and compensation structure reviews.
S1–17 Incidents, complaints and severe human rights impacts	Companies should disclose incidents and complaints related to human rights	Associated consulting services: Minimum Safeguard Criteria assessment, Human rights Due Diligence (HRDD) (See Sect 2.1.1), impact assessments, incident response planning.

Source: Author elaboration on the basis of EFRAG guidelines
[a]Diversity, Equity, and Inclusion strategy is a structured approach that organizations use to create a more diverse workforce, ensure fair treatment and access to opportunities, and foster an inclusive environment where all individuals feel valued and respected

stakeholders to understand the material impacts on value chain workers connected with the company's own operations and value chain, including through its products or services, as well as through its business relationships, and its related material risks and opportunities.

The Figure below reports disclosure requirements under ESRS S2 (Table 2.10).

The Topical ESRS S3 Standard Refers to affected Communities. The objective of this Standard is to specify disclosure requirements which allow to understand material impacts on affected communities connected with the company's own operations and value chain, including through its products or services, as well as through its business relationships, and its related material risks and opportunities.

The Table below reports disclosure requirements under ESRS S3 and the associated consulting services (Tables 2.11 and 2.12).

Table 2.10 Disclosure requirements for ESRS S2 and the supporting strategic consulting services

Disclosure requirement	Requirement description	Associated consulting services
S2–1: Policies related to value chain workers	The companies should describe the policies they have adopted to manage their material impacts on value chain workers, as well as associated material risks and opportunities. They have also to disclose human rights policies committments.	Associated consulting services: Policy development for responsible sourcing, human rights due diligence advisory.
S2–2: Processes for engaging with value chain workers about impacts	Companies should disclose their general processes for engaging with value chain workers and their representatives about actual and potential impacts on them.	Associated consulting services: Stakeholder engagement strategy (see Sect. 4.2 for methodologies for stakeholders mapping and engagement), supply chain communication frameworks, Survey design and data analytics.
S2–3: Processes to remediate negative impacts and channels for value chain workers to raise concerns	Companies should describe the processes they have in place to provide for or cooperate in the remediation of negative impacts on value chain workers that the undertaking is connected with, as well as channels available to value chain workers to raise concerns and have them addressed. They should also describe grievance mechanisms and remediation processes.	Associated consulting services: Grievance mechanism design, third-party hotline implementation[a], remediation planning
S2–4: Taking action on material impacts, risks, and opportunities	Companies should describe actions taken to address material impacts on value chain workers, and how risks and opportunities are managed.	Associated consulting services: ESG action planning, supply chain risk mitigation, impact remediation programs
S2–5: Tracking effectiveness of actions and approaches	Companies should disclose how the effectiveness of actions and approaches is monitored, including metrics and outcomes.	Associated consulting services: ESG assessment and performance monitoring, KPI development, impact evaluation and reporting systems

Source: Author elaboration on the basis of EFRAG Guidelines
[a]A third-party hotline implementation refers to setting up a confidential communication channel—usually managed by an external service provider—that allows workers (including those in the value chain) to report concerns, grievances, or misconduct safely and anonymously

In order to assess the impact of an action or a project on affected communities, it is possible to perform a Social Cost–Benefit Analysis in order to integrate social considerations in business practices and strategic decision-making. Chap. 7 provides a description of this kind of analysis which provides useful insights to communicate to stakeholders.

The Topical ESRS S4 standard refers to Consumers and Users. The objective of this Standard is to specify disclosure requirements which will enable to understand

Table 2.11 Disclosure requirements for ESRS S3 and the supporting strategic consulting services

Disclosure requirement	Requirement description	Associated consulting services
S3–1—Policies related to affected communities	Companies should disclose their policies to manage material impacts on affected communities, including human rights policies.	Associated Consulting Services: Community impact policy development, human rights due diligence advisory
S3–2—Processes for engaging with affected communities about impacts	Companies should describe how they engage with affected communities to understand and address actual or potential impacts.	Associated Consulting Services: Stakeholder engagement planning (see Sect. 4.2 for methodologies for stakeholders mapping and engagement), participatory consultation facilitation, Survey design and data analytics.
S3–3—Processes to remediate negative impacts and channels for affected communities to raise concerns	Companies should describe the processes it has in place to provide for or cooperate in the remediation of negative impacts on affected communities, as well as channels available to affected communities to raise concerns and have them addressed. They should also explain grievance mechanisms and remediation processes available to affected communities.	Associated Consulting Services: Grievance mechanism design, third-party mediation services, remediation strategy definition.
S3–4—Taking action on material impacts, risks, and opportunities	Companies should describe actions taken to address material impacts on affected communities and how risks and opportunities are managed.	Associated Consulting Services: Community development programs, risk mitigation planning, social investment strategy and actions definition (See Table 2.12 for an example of actions that can be taken to manage material impacts on affected communities).
S3–5—Tracking effectiveness of actions and approaches	Companies should disclose how the effectiveness of actions and approaches is monitored, including metrics, targets, and outcomes.	Associated Consulting Services: Impact measurement frameworks, ESG assessment and performance tracking, community feedback systems.

Source: Author elaboration on the basis of EFRAG Guidelines

Table 2.12 Example of actions that can be taken to manage material impacts on affected communities according to ESRS S3–4

Type of action	Example	Purpose
Community investment	Building schools, clinics, or clean water infrastructure in local communities	To address basic needs and improve quality of life
Environmental remediation	Cleaning up pollution or restoring ecosystems affected by company operations	To mitigate environmental harm to local populations
Cultural preservation	Supporting indigenous cultural heritage programs or protecting sacred sites	To respect and preserve cultural identity and rights
Health and safety programs	Providing health screenings, vaccinations, or safety training for nearby communities	To reduce health risks linked to company activities
Employment and training	Offering job training or local hiring initiatives	To create economic opportunities and reduce inequality
Consultation and consent	Conducting free, prior, and informed consent (FPIC) processes with indigenous groups	To ensure ethical and legal project development
Disaster preparedness	Supporting local emergency response plans or infrastructure	To reduce vulnerability to industrial or climate-related risks
Access to grievance mechanisms	Establishing local offices or digital platforms for complaints and feedback	To ensure communities can raise concerns and seek redress

Source: Author elaboration

material impacts on consumers and end-users connected with the company's own operations and value chain, including through its products or services, as well as through its business relationships, and its related material risks and opportunities.

The Table below reports disclosure requirements under ESRS S4, along with their descriptions and the associated consulting services (Tables 2.13 and 2.14).

2.2.3 Governance Dimension Topical ESRS

The last ESRS G1 standard regards Governance and in particular Business Conduct. The Standard focusses on:

- Business ethics and *corporate culture*, including anti-*corruption* and anti-*bribery*, the protection of whistleblowers, and animal welfare;
- The management of relationships with *suppliers*, including payment practices, especially with regard to late payment to small and medium-sized undertakings;
- Activities and commitments of the undertaking related to exerting its political influence, including its *lobbying activities.*

The Table below reports disclosure requirements under ESRS G1, along with their descriptions and the associated consulting services (Table 2.15).

It is worth underlining, however, that EFRAG is revising the first set of ESRS to reduce the reporting burden for in-scope entities with a deadline requested by

Table 2.13 Disclosure requirements for ESRS S4 and the supporting strategic consulting services

Disclosure requirement	Requirement description	Associated consulting services
S4–1—Policies related to consumers and end-users	The undertakings should disclose the adopted policies to manage material impacts on consumers and end-users, including those related to product safety, data privacy, and accessibility.	Consumer protection policy development, digital ethics and privacy advisory
S4–2—Processes for engaging with consumers and end-users about impacts	The undertakings should describe how they engage with consumers and end-users to understand and address actual or potential impacts.	Customer feedback systems, user research and co-creation workshops, customer satisfaction survey design.
S4–3—Processes to remediate negative impacts and channels for consumers and end-users to raise concerns	The undertakings should explain grievance mechanisms and remediation processes available to consumers and end-users.	Customer-complaint resolution systems, third-party mediation services
S4–4—Taking action on material impacts, risks, and opportunities	The undertakings have to describe actions taken to address material impacts on consumers and end-users and how risks and opportunities are managed (Table 2.14 provides some examples).	Product redesign for safety, digital accessibility upgrades, ethical AI implementation
S4–5—Tracking effectiveness of actions and approaches	The undertakings must disclose how the effectiveness of actions and approaches is monitored, including metrics, targets, and outcomes.	Customer satisfaction tracking, product safety audits, and digital trust metrics

Source: Author elaboration on the basis of EFRAG Guidelines

Table 2.14 Example of possible actions that can be taken to manage material impacts for ESRS S4-4

Type of action	Example	Purpose
Product safety enhancements	Redesigning products to eliminate hazardous materials or improve safety features	To reduce harm and liability, and protect consumer well-being
Digital accessibility improvements	Updating websites and apps to meet Web Content Accessibility Guidelines (WCAG) standards for users with disabilities	To ensure inclusive access and avoid discrimination
Data privacy protections	Implementing stronger encryption, consent mechanisms, and transparent data use policies	To protect user data and comply with GDPR or similar regulations
Ethical AI and algorithm design	Auditing algorithms for bias and fairness, and providing explainability tools	To build trust and prevent discriminatory outcomes
Sustainable product design	Using eco-friendly materials and designing for recyclability	To reduce environmental impact and meet consumer expectations

(continued)

Table 2.14 (continued)

Type of action	Example	Purpose
Consumer education campaigns	Launching initiatives to inform users about safe product use or digital literacy	To empower consumers and reduce misuse or harm
Inclusive product development	Co-creating products with underrepresented user groups	To ensure relevance, usability, and social equity
Complaint resolution improvements	Streamlining customer service and integrating AI chatbots for faster issue resolution	To enhance user satisfaction and responsiveness

Table 2.15 ESRS G1

Disclosure requirement	Requirement description	Associated consulting services
G1-1—Business conduct culture and policies	Disclose the undertaking's policies and culture related to ethical business conduct, including anti-corruption, anti-bribery, and fair competition	Ethics policy development, compliance training, anti-corruption program design
G1-2—Management of relationships with suppliers regarding business conduct	The company should provide information about the management of its relationships with its suppliers and its impacts on its supply chain	Supplier code of conduct implementation, third-party risk assessments
G1-3—Preventing and detecting corruption and bribery	Explain systems and controls in place to prevent, detect, and respond to corruption and bribery	Internal audit systems, whistleblower hotline setup
G1-4—Incidents of corruption or bribery	The undertaking should provide information on incidents of corruption or bribery during the reporting period	The consultants can support the clients in the reporting activity
G1-5—Political influence and Lobbying activities	Disclose the undertaking's lobbying practices, including objectives, expenditures, and alignment with public commitments	Transparency reporting, lobbying policy development, stakeholder alignment reviews
G1-6—Payment practices	The undertaking should provide information on its payment practices, especially with respect to late payments to small and medium enterprises (SMEs)	The consultants can support the clients in the reporting activity

the EC of 31 October 2025. The Exposure Draft outlining the proposed revisions has been published and a public consultation period has been started (The public consultation runs from 31 July to 29 September 2025). In the ESRS Revised version, the number of mandatory datapoints—to be reported if material—has been reduced by 57%, while the total volume of disclosures, including both mandatory and voluntary items, has been cut by 68%. Overall, the ESRS standards have been shortened by more than 55%, significantly enhancing their accessibility and ease of implementation, particularly for companies newly falling under the CSRD scope beyond those already reporting for the 2024 financial year.[48]

[48] https://www.efrag.org/en/news-and-calendar/news/press-release-efrag-shares-revised-esrs-exposure-drafts-and-launches-60day-public-consultation

With this revision, the European Commission aims to alleviate unnecessary administrative burdens, while still meeting the core policy objectives of the European Green Deal.[49]

Focus

The CSRD Implementation, Evidence from the 2025 Reporting Cycle

In 2025, companies across Europe have published, for the first time, disclosures aligned with the Corporate Sustainability Reporting Directive (CSRD) and the European Sustainability Reporting Standards (ESRS). 11,000 companies are expected to report by the end of the year.[50]

A Survey conducted by Datamaran[51,52] between December 2024 and January 2025, shows how companies see the identification of material impacts, risks, and opportunities (IROs) as the most valuable element of CSRD reporting for internal decision-making. Through IROs identification, companies developed a new tool to manage environmental, social, and governance (ESG)-related risks and protect their long-term value.

The Survey results show that negative impacts dominated CSRD disclosures; while social topics like Own Workforce (S1) and Consumers & End-Users (S4) had the highest share of positive impacts. Opportunities overall remain underrepresented, this may indicate that companies are still developing the capabilities to identify and articulate how sustainability initiatives can unlock long-term value, innovation, growth, and resilience.

Almost all companies reported on Climate Change, Own Workforce, and Business Conduct, signaling these as core sustainability priorities. In contrast, Biodiversity, Water, and Affected Communities were the least reported topics.[53]

Overall, the survey results reveal a growing commitment among businesses to sustainability, with the CSRD acting as a catalyst for accelerated planning and improved ESG integration. However, while companies recognize the importance of assessing impact, risks, opportunities; strategic alignment, and governance, implementation remains inconsistent—particularly in areas such as value chain integration, data analytics, and assurance preparedness.

[49] Brussels, 26.2.2025 COM(2025) 80 final 2025/0044 (COD) "amending Directives (EU) 2022/2464 and (EU) 2024/1760 as regards the dates from which Member States are to apply certain corporate sustainability reporting and due diligence requirements"

[50] Datamaran (2025) "CSRD Reports Uncovered: Benchmarking the First Wave of 2025 Disclosures"

[51] The Survey gathered insights from professionals at over 65 companies; it is based on a review of 304 CSRD sustainability statements and 11,208 individual IRO statements from companies across 21 countries and 11 broad business sectors spanning 57 industries.

[52] Datamaran (2025) "2025 CSRD Pulse Check Survey: Key Insights from Companies on Value, Engagement, and Challenges"

[53] Datamaran (2025) "CSRD Reports Uncovered: Benchmarking the First Wave of 2025 Disclosures", pg. 5.

2.3 The International Sustainability Reporting Frameworks

Historically, sustainability reporting has suffered from a lack of consistency and standardization. Companies often relied on different frameworks and methodologies, which made it difficult to compare data across organizations and sectors. This fragmentation not only hindered transparency but also undermined the confidence of investors and other stakeholders in the reliability and credibility of the disclosed information. As a result, the effectiveness of sustainability reporting as a tool for informed decision-making and long-term value assessment was significantly diminished.

Key international frameworks for ESG (Environmental, Social, and Governance) disclosure and sustainability reporting guide organizations in reporting their sustainability performance, ensuring transparency, and meeting stakeholder expectations (Table 2.16).

In the previous chapters, we analyzed CSRD and the bundle of companies for which it is mandatory in Europe the sustainability reporting according to ESRS. In this chapter, we will analyze the most significant international sustainability-related reporting regulations and standards, i.e. the International Financial Reporting Standards for sustainability disclosures (IFRS), the Global Reporting Initiatives (GRI) and the SEC's Climate Disclosure Rule and the State of California sustainability-related reporting regulations and standards which are the most frequently adopted as voluntary frameworks but are also mandatory for some Jurisdictions (Fig. 2.5).

Table 2.16 Key features of sustainability reporting Source: Filho et Al. 2025[a]

Key features of sustainability reporting	
Transparency	Providing stakeholders with a transparent view of an organization's sustainability performance.
Performance improvement	Encouraging organizations to improve their sustainability performance over time.
Value creation	Demonstrating how sustainability practices contribute to the long-term value creation of the organization.
Accountability	Holding organizations accountable for their impacts on society and the environment
Stakeholder engagement	Engaging with stakeholders and responding to their concerns and expectations regarding sustainability.
Reputation	Enhanced reputation and brand value.

[a]Leal Filho, W., Wall, T., Williams, K., Dinis, M. A. P., Fernandez Martin, R. M., Mazhar, M., & Gatto, A. (2025). European sustainability reporting standards: An assessment of requirements and preparedness of EU companies. Journal of Environmental Management, 380, https://doi.org/10.1016/j.jenvman.2025.125008

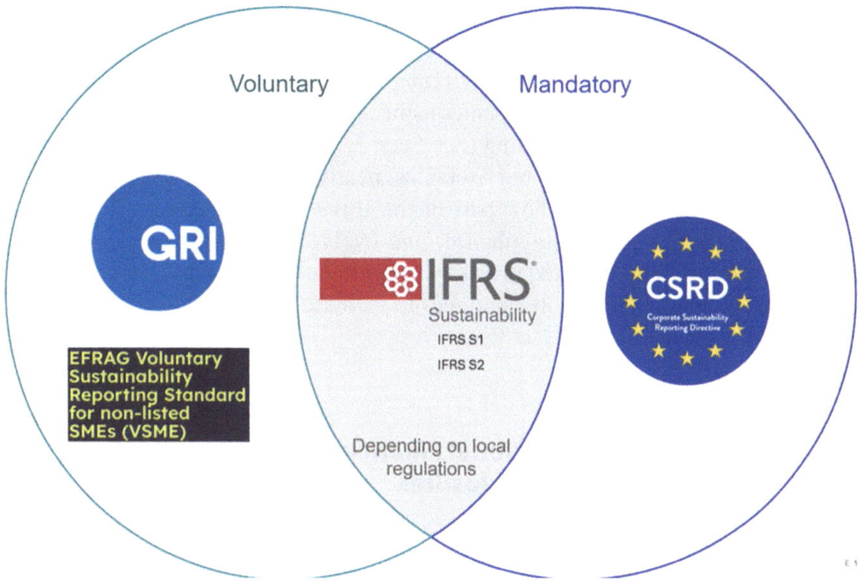

Fig. 2.5 Main ESG reporting frameworks today. Source: Bureau Veritas. Notes: An approach is mandatory or voluntary according to the jurisdictions

FOCUS: Some Data on Sustainability Reporting in the World

According to the KPMG Survey of Sustainability Reporting 2024,[54] today 96% of G250 companies[55] report on sustainability and 56% of them have a sustainability leader.

The Survey identified some major trends:

1. Reporting on sustainability and setting carbon targets has become part of business as usual (95% of G250 companies publish carbon targets (up 15 percentage points since 2022).
2. Some companies have already changed practices in advance of the move to mandatory reporting on sustainability under the EU's CSRD.
3. Double materiality (See Sect. 4.3), required under CSRD, is now used by half of the largest companies.

[54] KPMG (2024) The move to mandatory reporting—Survey of Sustainability Reporting 2024. KPMG International

[55] The world's 250 largest companies by revenue based on the 2023 Fortune 500 ranking

4. Despite moves toward mandatory reporting, voluntary guidelines and standards remain widely used (GRI remains the most popular standard, with three-quarters of G250 companies using it; See Sect. 3.2.2 for details. Also the use of SABS is increasing (See Sect. 3.2.5). SASB Standards are used by 67% of companies in the Americas, nearly twice the proportion in any other region. SASB is now part of the IFRS Foundation. The ISSB has committed to maintain, enhance, and evolve the SASB Standards and encourages preparers and investors to continue to use them.
5. Reporting on biodiversity continues to increase.

2.3.1 ISSB/IFRS—International Financial Reporting Standards for Sustainability Disclosures

The International Financial Reporting Standards (IFRS) Foundation announced the formation of the **International Sustainability Standards Board (ISSB)** on 3 November 2021 at COP26 in Glasgow.

The ISSB has set out four key objectives:

- To develop standards for a global baseline of sustainability disclosures.
- To meet the information needs of investors.
- To enable companies to provide comprehensive sustainability information to global capital markets.
- To facilitate interoperability with disclosures that are jurisdiction-specific and/or aimed at broader stakeholder groups.

The International Sustainability Standards Board (ISSB) is, thus, developing a high-quality, comprehensive global baseline of sustainability disclosures focused on the needs of investors and the financial markets.[56]

ISSB's first two IFRS Sustainability Standards were issued in June 2023,[57] IFRS S1 and IFRS S2, which catalyze the integration of sustainability with financial reporting. Furthermore, the ISSB has developed application guidance, along with illustrative guidance and illustrative examples to support companies in applying IFRS S1 and IFRS S2.

These Standards are mandatory for some Jurisdictions (the different phases of adoption by country are reported in Fig. 2.6) and voluntary in others.

In order to foster a further adoption of these standards, the ISSB is committed to continue supporting their implementation by developing further guidance and

[56] The ISSB builds on the work of market-led investor-focused reporting initiatives, including the Climate Disclosure Standards Board (CDSB), the Task Force for Climate-related Financial Disclosures (TCFD), the Value Reporting Foundation's Integrated Reporting Framework and industry-based SASB Standards, as well as the World Economic Forum's Stakeholder Capitalism Metrics.

[57] IFRS S1 and IFRS S2 are effective from January the first 2024.

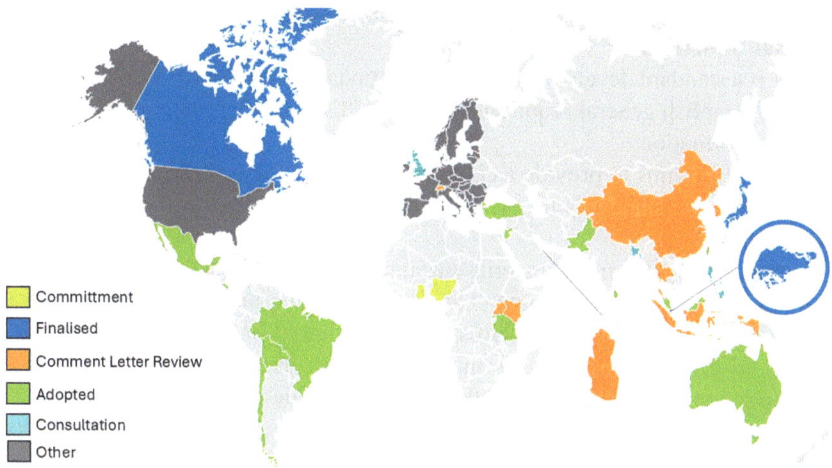

Fig. 2.6 Phases of mandatory adoption of IFRS Sustainability Disclosure Standards by Jurisdictions. Source: Deloitte (2025) Adoption of IFRS Sustainability Disclosure Standards by Jurisdictions

training material, and establishing a Transition Implementation Group on IFRS S1 and IFRS S2 (TIG).

It is worth to underline that the International Sustainability Standards Board (ISSB) and the European Commission, in collaboration with EFRAG, have worked in close coordination to ensure strong interoperability between the European Sustainability Reporting Standards (ESRS) and the IFRS Sustainability Disclosure Standards (ISSB Standards). This collaboration has prioritized coherence in climate-related disclosures, a critical area for sustainability practitioners.

Key outcomes include:

- **Alignment on financial materiality**: The ESRS adopts a definition of financial materiality that is consistent with IFRS S1 *General Requirements for Disclosure of Sustainability-related Financial Information,* supporting comparability across jurisdictions.
- **Harmonized terminology**: Both frameworks utilize a shared set of defined terms, facilitating consistent interpretation and application.
- **Converged climate disclosures**: There is a high degree of alignment in climate-related reporting requirements.[58]

[58] EFRAG & IFRS, 2024. ESRS-ISSB standards. Interoperability guidance. Available from: https://www.efrag.org/Assets/Download?assetUrl=/sites/webpublishing/SiteAssets/ESRS-ISSB+Standards+Interoperability+Guidance.pdf

IFRS S1—General Requirements for Disclosure of Sustainability-Related Financial Information

IFRS S1 is a standard developed by the International Sustainability Standards Board (ISSB) to establish general requirements for the disclosure of sustainability-related financial information.

The standard aims to provide a comprehensive global baseline of sustainability-related disclosure standards to meet the information needs of investors and other stakeholders.

It covers the disclosure of material sustainability-related risks and opportunities, including those related to climate change, biodiversity, and other environmental, social, and governance (ESG) factors. This is required because the capacity of a company to generate cash-flow is interrelated with its capacity to manage risks related to society, environment, and the economy through the entire value chain.

The Standard, thus, provides a guidance to companies to perform sustainability-related financial disclosures in order to clearly identify risks and opportunities that could affect their cash flows, access to finance, and cost of capital over the short, medium, and long term.

It is interesting to focus on the definition of materiality provided in IFRS S1: "*In the context of sustainability-related financial disclosures, information is material if omitting, misstating or obscuring that information could reasonably be expected to influence decisions that primary users of general purpose financial reports make on the basis of those reports, which include financial statements and sustainability-related financial disclosures and which provide information about a specific reporting entity.*"[59] It is clear how the final purpose of the Standard is to provide the stakeholders all the relevant information on the base of which they take decisions relating to providing resources to the entity.

In particular, an entity shall provide disclosures about:

- Governance processes, controls, and procedures used to monitor and manage sustainability-related impacts, risks, and opportunities.
- Strategy adopted to manage sustainability-related impacts, risks, and opportunities; the resilience of the entity's strategy and its business model to those sustainability-related risks
- The risk management process, specifying the time horizon over which the effects of the sustainability-related risks and opportunities could reasonably occur.
- Metrics and targets to measure the entity's performance.

The sustainability reporting shall cover the same entities as the financial reporting for consistency and it has to be published at the same time as the financial statements. Data have to be reported also for the previous period to allow comparability (this applies since the second year of reporting).

[59] IFRS S1 "General Requirements for Disclosure of Sustainability-related Financial Information" pg 8

IFRS S1 requires an entity to refer to and consider the applicability of specified sources of guidance in preparing its sustainability-related financial disclosures. In particular, it mentions SABS Standards and the Climate Disclosure Standards Board (CDSB) Framework Application Guidance.

The Sustainability Accounting Standards Board (SASB) is today part of the IFRS Foundation, forming part of the International Sustainability Standards Board (ISSB). The ISSB has committed to maintain, enhance, and evolve the SASB Standards and encourages preparers and investors to continue to use them. SASB Standards are used by 67% of companies in the Americas, nearly twice the proportion in any other region and is available for 77 industries identifying the sustainability-related risks and opportunities.

The SASB Standards are organized by industry and the disclosure topics describe specific sustainability-related risks and opportunities associated with the activities conducted by entities within a particular industry.[60] In some cases, an entity might conclude that a disclosure topic would not result in useful information because of the entity's business model; in other cases, the entity may need to add some specific disclosures.

SASB organizes sustainability topics into five dimensions:

1. Environment
2. Human Capital
3. Social Capital
4. Business Model and Innovation
5. Leadership and Governance

CDSB Framework Application Guidance is another guidance mentioned by IFRS S1 to identify sustainability-related risks and opportunities that could reasonably be expected to affect the entity's prospects. In particular, The CDSB Framework Application Guidance[61] can support entities in identifying water- and biodiversity-related risks and opportunities.

In summary, IFRS S1 is structured to ensure the disclosure of sustainability-related information that complements an entity's financial statements, irrespective of the specific Accounting Principles applied. This approach promotes consistency and comparability across entities, enhancing the usefulness of financial reporting for investors and other stakeholders.

IFRS S2—Climate-Related Disclosures

IFRS S2 is a standard developed by the ISSB to establish requirements for the disclosure of climate-related risks and opportunities. The standard aims to provide investors and other stakeholders with information about an organization's exposure to climate-related risks and opportunities, and how these are managed.

---

[60] IFRS S1 "Accompanying Guidance on General Requirements for Disclosure of Sustainability-related Financial Information", pg. 7

[61] The Climate Disclosure Standards Board (CDSB) is an international consortium of business and environmental NGOs committed to advancing and aligning the global mainstream corporate reporting model to equate natural capital with financial capital. It offers companies a framework for reporting environmental information with the same rigor as financial information.

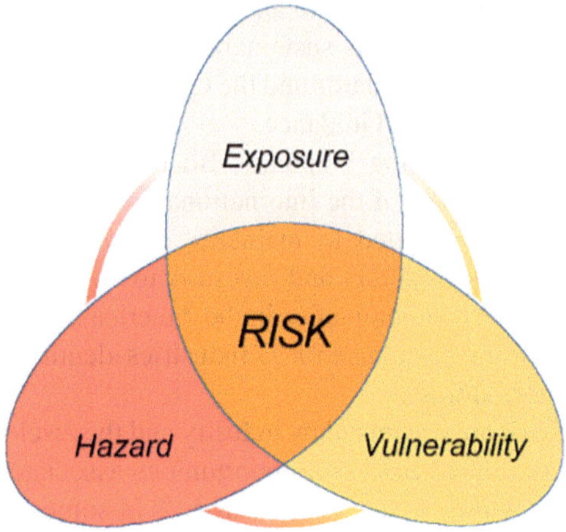

Fig. 2.7 Definition of risk. Source: Bureau Veritas

Climate change, in fact, brings significant risks to buildings and infrastructure, requiring a comprehensive approach to assess and address these challenges. Broadly speaking, the assessment of a specific risk is given by the intersection of Hazard,[62] Exposure, and Vulnerability[63] (See: Fig. 2.7). Therefore, in risk assessment, all these aspects have to be taken into account.

Given these premises, considering climate-related risks, it is worth underlining that they fall into two categories:

- **Physical Risks**: Risks an entity faces due to climate change. These can be acute, such as extreme weather events like floods, wildfires, hurricanes, and heatwaves, which cause immediate and often severe damage. They can also be chronic, involving long-term shifts in climate patterns, such as rising sea levels, desertification, and changing precipitation trends, all of which can disrupt supply chains, damage assets, and threaten human health and safety.
- **Transition Risks**: Risks an entity faces due to the transition toward a low-carbon economy. These include policy and legal changes, such as the introduction of carbon taxes or stricter environmental regulations, as well as technological risks linked to the rapid development of clean technologies that may render existing systems obsolete. Market risks also play a role, as consumer preferences evolve and commodity prices fluctuate, potentially leading to asset devaluation. Additionally, reputational risks can emerge if companies fail to meet environmental expectations, resulting in a loss of investor confidence and customer trust.

The standard IFRS S2 requires organizations to disclose information about their transition plans, scenario analysis, and the financial impact of climate-related risks

[62] Examples of Hazards are Flooding, Heatwave, Storms, etc.

[63] Vulnerability can be physical, operational, and organizational

and opportunities. It also provides guidance on the disclosure of Scope 1, 2, and 3 greenhouse gas emissions (GHG) generated during the reporting period, expressed as metric tons of CO_2 equivalent (See Sect. 6.2 for details on GHG calculations), as well as the use of carbon pricing and carbon offsets. In particular, the company shall indicate if it has planned the use of *carbon credits* to offset greenhouse gas emissions to achieve any net greenhouse gas emissions target (See Sect. 6.2.3 for details about the concept of carbon credits). It is important to describe the governance processes, controls and procedures used, and the strategy adopted by the company to monitor and manage climate-related risks and opportunities. This usually results in a climate-related scenario analysis and/or a resilience and adaptation action plan. **The Adaptation Action Plan** is usually developed by companies with the support of consultants and it serves as a comprehensive roadmap for the operator to systematically address the climate-related risks and vulnerabilities identified in the assessment. The disclosure of climate-related risks and opportunities shall enable users to understand their effects on the company's business model, value chain, financial performance, and cash flows with an *outside-in* point of view.

A general procedure that can be followed to identify and prioritize climate-related hazards, assess the vulnerability of buildings and infrastructure, and develop a resilience and adaptation action plan is composed of three-steps.

The first step involves the identification and prioritization of relevant climate hazards. This includes analyzing historical data, climate projections, and site-specific factors to determine the likelihood and potential impact of hazards, such as extreme weather events, temperature fluctuations, and changes in precipitation patterns. By prioritizing the most significant hazards, stakeholders can focus their efforts on the most pressing risks.

The second step involves conducting a vulnerability assessment to evaluate how susceptible the building or infrastructure is to the identified climate hazards. This process includes analyzing the structural soundness, building systems, and operational procedures to assess their capacity to withstand these threats. Key considerations include the building's age, architectural design, and maintenance record, as well as the risk of cascading failures and dependencies on external infrastructure.

The final step focuses on formulating a comprehensive **Resilience and Adaptation Action Plan**. This plan defines targeted strategies to strengthen the building/infrastructure's capacity to endure, respond to, and recover from climate-related disruptions. It encompasses a range of recommendations, including structural upgrades, enhancements to operational protocols, emergency preparedness initiatives, and coordinated efforts with local authorities and community stakeholders. The plan also considers long-term sustainability goals, aiming to future-proof the building against evolving climate risks while supporting continuity of use, safety, and environmental performance.

In order to define a tailored Resilience and Adaptation Action Plan, it becomes relevant to understand the initial Company's Climate Risk Maturity Level. To this aim, Aon[64], for instance, has defined a model which allows companies to assess

[64] Aon (2022) Assessing Your Company's Climate Maturity: A Three-Step Strategy to Address New Climate Regulations and Disclosure Standards. Available at: https://humancapital.aon.com

Fig. 2.8 Climate risk maturity model. Source: Aon (2022) Assessing Your Company's Climate Maturity: A Three-Step Strategy to Address New Climate Regulations and Disclosure Standards

where they currently stand on their journey toward a robust climate strategy (See Figure above) identifying practical steps to progress that could make easier to define effective strategies (Fig. 2.8).

The framework above provides a structured roadmap for evolving from basic environmental awareness, to fully embedding climate risk into business strategy and governance.

In the initial stages, companies often engage in standalone sustainability efforts, track basic environmental indicators, and assign preliminary responsibilities. While these steps are foundational, they fall short of meeting increasing regulatory demands and stakeholder expectations.

As organizations progress, they begin to implement board-level oversight, define clear environmental objectives, and carry out formal climate risk assessments. This includes leveraging scenario analysis tools and tracking climate trends to better align with strategic goals.

With growing maturity, climate risks become integrated into enterprise risk management systems. Companies start reporting consistently, adopt global standards, and set science-based targets that support transition pathways.

At advanced stages, firms evaluate climate risks across their entire value chain and distribute accountability across business units. Leading organizations also tie executive compensation to environmental performance, reinforcing top-level accountability.

The model emphasizes the critical role of both internal and external communication throughout this journey. Transparent engagement fosters alignment, enhances reporting quality, and supports informed decision-making. By identifying their position on this maturity curve, companies can focus on targeted improvements,

Focus

How to Report Sustainability Risks and Opportunities According to IFRS S1 and IFRS S2

In order to prepare a Sustainability Report according to IFRS S1 and IFRS S2, the company shall disclose all the required information explicitly indicating, within each Chapter, the referring standard and the paragraph number the disclosed information refers to.

For example, if the company describes its business model and the value chain, the reference will be IFRS S1.32. IFRS S1.32 refers to paragraph 32 of the IFRS S1—General Requirements for Disclosure of Sustainability-related Financial Information, which states that "*An entity shall disclose information that enables users of general purpose financial reports to understand the current and anticipated effects of sustainability-related risks and opportunities on the entity's business model and value chain. Specifically, the entity shall disclose: (a) a description of the current and anticipated effects of sustainability related risks and opportunities on the entity's business model and value chain; and (b) a description of where in the entity's business model and value chain sustainability-related risks and opportunities are concentrated (for example, geographical areas, facilities and types of assets).*"

The Chapters usually contained in a sustainability report prepared according to the IFRS Sustainability Disclosure Standards are the following[65]:

- **Basis of Preparation**: Where it is reported a declaration of compliance with IFRS Sustainability Disclosure Standards and the connection with financial statements (reporting period, reporting entity, and presentation currency). It has to be indicated the time of adoption of IFRS Sustainability Disclosure Standards and transitional provisions.
- **Overview of the group and value chain:** It shall describe the group's primary and secondary business activities, corporate structure, geographical locations, strategy and sustainability-related goals and the value chain.
- **Reporting boundary:** Description of the reporting entity and its assets, description of the boundaries for GHG emissions calculations.
- **Judgements and measurement uncertainties:** Information about the materiality process and metrics with a high degree of measurement uncertainty.
- **Materiality assessment:** Identification of sustainability-related risks and opportunities, identification of material information, outcome of the process with a list of the risks identified, interaction with material information in the financial statements.

[65] PwC. (2025). VALUE Plc—Example sustainability disclosures under IFRS Sustainability Disclosure Standards.

- **Sustainability governance:** Description of the structure of the governance body with a representation of the organizational chart, description of the management's role in governance and the impact of sustainability on remuneration policies, description of the risk management process.
- **Environment-related risks and opportunities:** Description of the climate-related physical risks the entity is exposed to. Their effects on business model, value chain, strategy and the decision-making process, and the company's capacity to adapt. Description of the current and anticipated financial effects. Also a climate-related scenario analysis has to be performed to understand and evaluate the potential impacts of the identified climate-related risk on the group's operations, using a range of expected scenarios. It is also important to describe processes, controls, and policies to manage climate-related risks and opportunities; and metrics and the targets used to monitor performance and their achievement.
- **Social-related risks and opportunities**: The analysis of these risks and opportunities should be done as per the environmental ones described above.
- **Governance-related risks and opportunities**: The analysis of these risks and opportunities should be done as per the environmental ones described above.
- **Events after the reporting period**: It refers to IFRS S1.68 which states that "*an entity shall disclose information about transactions, other events and conditions that occur after the end of the reporting period, but before the date on which the sustainability-related financial disclosures are authorized for issue, if non-disclosure of that information could reasonably be expected to influence decisions that primary users of general purpose financial reports make on the basis of those reports.*"

enhance the credibility of their climate risk management, and better align ESG initiatives with long-term business resilience.

2.3.2 GRI—Global Reporting Initiative (GRI) Standards

The Global Reporting Initiative (GRI) is an independent, international organization that helps businesses, governments, and other organizations understand and communicate their impacts on critical issues such as climate change, human rights, and corruption.

The GRI Standards are one of the voluntary most widely used and comprehensive sustainability reporting frameworks globally. They cover a wide range of environmental, social, and governance (ESG) topics and help organizations report their sustainability performance in a standardized manner.

GRI Standards are organized in a modular way. In particular, they are divided into:

- Universal Standards (See Fig. 2.9).

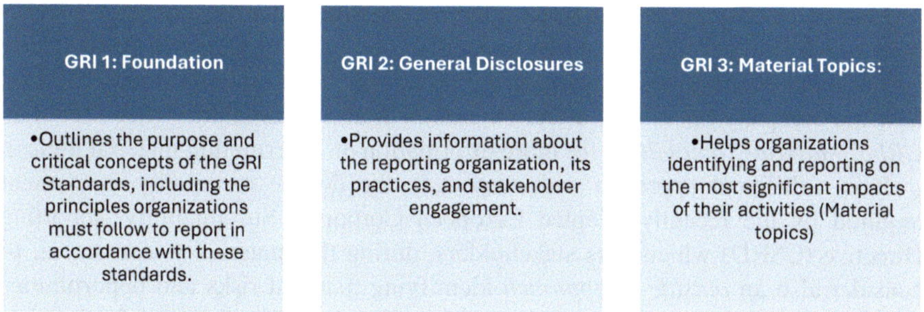

Fig. 2.9 GRI Universal Standards

- Sector Standard: Provide specific guidelines for different sectors, ensuring that the reporting is relevant and consistent across similar organizations.
- Topic Standards: These standards cover specific topics such as emissions, waste, labor practices, and human rights. They include detailed disclosures that organizations can use to report on these areas.

The key elements that lay the foundation for sustainability reporting by using GRI Standards are[66]:

- Impacts;
- Material Topics;
- Due Diligence;
- Stakeholders.

"Impact refers to the effect an organization has or could have on the economy, environment, and people, including effects on their human rights, as a result of the organization's activities or business relationships. The impacts can be actual or potential, negative or positive, short-term or long-term, intended or unintended, and reversible or irreversible. These impacts indicate the organization's contribution, negative or positive, to sustainable development. [...] The impacts on the economy, environment, and people are interrelated."[67]

Most significant impacts on the economy, environment, and people, defined as material for the organization, have to be reported. The GRI Standards group such impacts into topics, to help organizations report in a standardized way about multiple impacts that relate to the same topic.

[66] https://www.globalreporting.org/publications/documents/english/gri-1-foundation-2021/
[67] GRI1 -Foundation 2021

Materiality assessment is a structured process which involves company's stakeholders who are asked to prioritize impacts on the basis of their relevance, taking into account an *inside-out approach*, i.e. assessing from multiple points of view, how the company operations affect environment and communities (See sect. 1 in *GRI 3: Material Topics 2021* for more information on determining material topics).

It is a different approach with respect to the double materiality assessment required by the recently adopted European Corporate Sustainability Reporting Directive (CSRD) which asks stakeholders, during the materiality assessment, to consider also an *outside-in approach* identifying material risks and opportunities which could affect the company's financial performance (See Sect. 4.1 for details).

Due diligence refers to the process by which an organization identifies, prevents, mitigates its actual and potential negative impacts on the economy, environment, and people. Due diligence focuses on identifying stakeholders' interests that are or could be negatively affected by the organization's activities. " *Common categories of stakeholders for organizations are business partners, civil society organizations, consumers, customers, employees and other workers, governments, local communities, non-governmental organizations, shareholders and other investors, suppliers, trade unions, and vulnerable groups.*"[68]

Reporting in accordance with the GRI Standards requires to meet the following Requirements[69]:

- Requirement 1: Apply the reporting principles;
- Requirement 2: Report the disclosures in GRI 2: General Disclosures 2021;
- Requirement 3: Determine material topics;
- Requirement 4: Report the disclosures in GRI 3: Material Topics 2021;
- Requirement 5: Report disclosures from the GRI Topic Standards for each material topic;
- Requirement 6: Provide reasons for omission for disclosures and requirements that the organization cannot comply with;
- Requirement 7: Publish a GRI content index;
- Requirement 8: Provide a statement of use;
- Requirement 9: Notify GRI.

In Fig. 2.10, it is reported the GRI Standards reporting process. After the understanding of the system and the key elements of the GRI Standards, the first requirement is to apply the reporting principles (Requirement 1). The reporting principles are fundamental to achieving high-quality sustainability reporting and can be summarized in the following[70]:

---

[68] GRI1 -Foundation 2021

[69] GRI1—Foundation 2021

[70] GRI 1—Foundation—Section 4

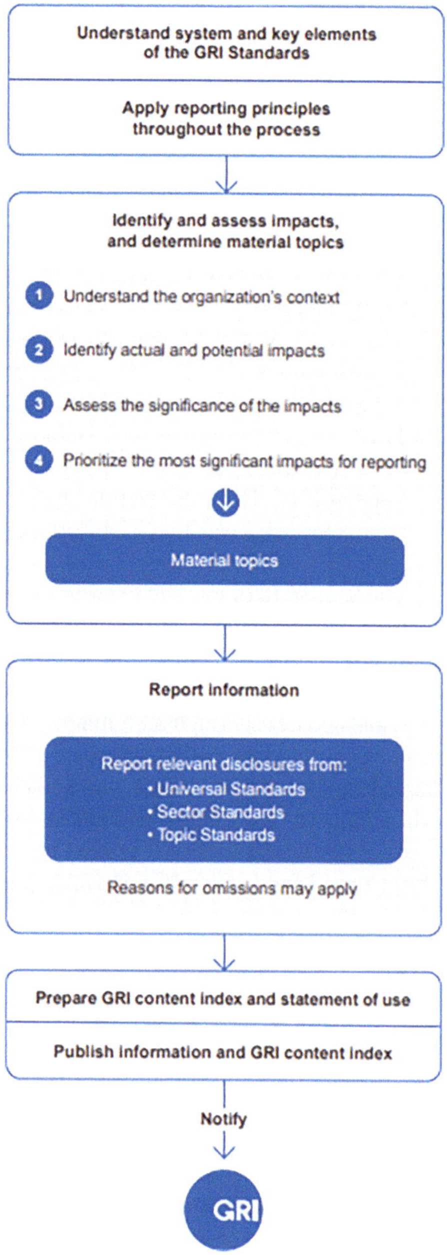

Fig. 2.10 Reporting with GRI Standards. Source: Global Reporting Initiative. (2021). A short introduction to the GRI Standards. https://www.globalreporting.org

- **Accuracy:** The organization shall report information that is correct and sufficiently detailed to allow an assessment of the organization's impacts.
- **Balance:** The organization shall report information in an unbiased way and provide a fair representation of the organization's negative and positive impacts.
- **Clarity:** The organization shall present information in a way that is accessible and understandable.
- **Comparability:** The organization shall select, compile, and report information consistently to enable an analysis of changes in the organization's impacts over time and an analysis of these impacts relative to those of other organizations.
- **Completeness:** The organization shall provide sufficient information to enable an assessment of the organization's impacts during the reporting period.
- **Sustainability context:** The organization shall report information about its impacts in the wider context of sustainable development.
- **Timeliness:** The organization shall report information on a regular schedule and make it available in time for information users to make decisions.
- **Verifiability**: The organization shall gather, record, compile, and analyze information in such a way that the information can be examined to establish its quality.

The Second Requirement regards the reporting of the General disclosures in GRI 2 about the organization, its strategies and its governance. After that it is necessary to identify impacts and assess material topics (Requirements 3, 4, and 5).

The materiality assessment is the first step and it starts with an analysis of the organization's context through an initial high-level overview of its activities, its business relationships and the sustainability context in which these occur, and an overview of its stakeholders (the organization should draw a full list of individuals and groups whose interests are affected or could be affected by the organization's activities). The analysis of the organization's context approach is similar to the business modeling one (See Chap. 7 for details). It is, in fact, necessary to understand the organization's purpose, value, or mission statements, the types of activities it carries out, main products and/or services and the business relationships in its value chain.

The second step is the identification of actual and potential impacts. This can be done considering the topics and impacts described in the Sector Standards.

The materiality assessment allows the evaluation of the significance of the impacts and their prioritization. Prioritization of the identified impacts allows the identification of the most significant ones for reporting. Material topics are often represented in a materiality matrix which reports on the *x-axes* the magnitude of the impact and on the *y-side* the importance to stakeholders.

Information on the material topics must be disclosed through GRI standards in a report that has to specify, in a GRI content Index, the GRI Standards that the organization has used. Reports using the GRI Standards may be published in various formats (e.g., electronic, paper-based) and made accessible. The content index makes reported information traceable and increases the report's credibility and transparency and the organization must specify in the content index if a "reason for omission" is being used (Requirement 6). In addition, the disclosure or the requirement that the organization cannot comply with, together with an explanation.

The GRI content Index (Requirement 7) and the statement of use (Requirement 8) have to be published and notified to GRI (Requirement 9) by sending an email to reported email address.[71]

The Statement of Use (Requirement 8) in the GRI 1 Foundation 2021 is an essential part of the reporting principles. It ensures that organizations clearly communicate how they are using the GRI Standards in their sustainability reports.

2.3.3 SEC's Climate Disclosure Rule and the State of California Sustainability-Related Reporting Regulations and Standards

This chapter will introduce the Securities and Exchange Commission (SEC) Climate Disclosure Rule and California's Climate Disclosure Laws (SB 253, SB 261, and AB 1305). Both aim to enhance transparency around climate-related risks and emissions and they are the mainly adopted sustainability reporting frameworks in Americas Region, together with IFRS S1 and S2.

The SEC's Climate Disclosure Rule,[72] adopted in March 2024, requires U.S. public companies to disclose material climate-related risks and the financial impacts of such risks. Investors view the regulations as a crucial move toward greater transparency and accountability, especially as climate-related risks increasingly influence corporate performance. As of March 27, 2025, even if it was adopted, it is now paused and, due to legal and political developments, it is not currently mandatory. It has, in fact, faced significant legal challenges from states and private parties, leading to a stay on its implementation. Its future depends on the outcome of the ongoing litigation and potential actions by future administrations.

The original rule's requirements are, however, more limited compared to the CSRD and California's regulations, e.g. the computation of Scope 3 GHG emissions is excluded from the SEC's disclosures. The rule, which is still under judicial review, will apply to larger registrants starting in 2026.

Key components are:

---

[71] As of 2025, the email address is: reportregistration@globalreporting.org

[72] SEC Final Rule Release No. 33–11,275, *The Enhancement and Standardization of Climate-Related Disclosures for Investors*.

- The disclosure of the impact of climate-related risks on the company's business model and the financial performance;
- Disclosure of GHG emissions;
- Disclosure of the Governance board to manage climate-related risks;
- Description of the risk of severe climate events and their impact on the business.

In 2023, California passed its **Climate Disclosure Laws** consisting of three major climate disclosure bills: **SB 253; SB 261, and AB 1305.**

SB 253 also called "Climate Corporate Data Accountability Act"[73] requires specified partnerships, corporations, limited liability companies, and other business entities with total annual revenues over $1 billion and that do business in California, defined as "reporting entities," to publicly disclose to the Scope 1 and Scope 2 GHG emissions starting in 2026 and begin reporting Scope 3 GHG emissions in 2027, on the basis of 2026 GHG emissions data. Reports must follow the Greenhouse Gas Protocol (GHGP)[74] and be verified by an independent third party.

SB 261[75] **also called "Climate-Related Financial Risk Act,"** mandates "covered entities"[76] to prepare a climate-related financial risk report disclosing the entity's climate-related financial risks and measures adopted to reduce and adapt to climate-related financial risks. A copy of the report should be published on the entity's Internet website.

Finally, the third bill is the **AB 1305 also called "Voluntary Carbon Market Disclosures Act"**[77] which requires covered business entities disclose specified information regarding the sale and use of certain voluntary carbon offsets (VCOs), as well as claims of net zero, carbon neutrality, or significant greenhouse gas ("GHG") emissions reductions. In particular, the corporations which market or sell VCOs within California, or operate in California and make public claims about reducing greenhouse gas (GHG) emissions, regardless of whether they actually purchase or use VCOs, are subjected to this law.

---

[73] California State Legislature. (2023). Senate Bill No. 253: Climate Corporate Data Accountability Act. Retrieved from https://leginfo.legislature.ca.gov/faces/billNavClient.xhtml?bill_id=202320240SB253

[74] See Sect. 6.2

[75] California State Legislature. (2023). *Senate Bill No. 261: Greenhouse gases: climate-related financial risk*. Retrieved from https://leginfo.legislature.ca.gov/faces/billNavClient.xhtml?bill_id=202320240SB261

[76] Covered entities are corporation, partnership, limited liability company, or other business entity formed under the laws of the state, the laws of any other state of the United States or the District of Columbia, or under an act of the Congress of the United States with total annual revenues in excess of five hundred million United States dollars ($500,000,000) and that does business in California

[77] California State Legislature. (2023). Assembly Bill No. 1305: Voluntary Carbon Market Disclosures Act. Retrieved from https://leginfo.legislature.ca.gov/faces/billPdf.xhtml?bill_id=202320240AB1305&version=20230AB130593ENR

It is important to underline that certain requirements of SB 253 and SB 261 were amended by SB-219 also called **"Greenhouse gases: climate corporate accountability: climate-related financial risk"** signed into law on September 27, 2024. It specifies that the reporting entities are authorized to consolidate emission disclosures at the parent-company level and gives California Air Resource Board (CARB) the discretion to establish a schedule for an entity's submission of Scope 3 emission disclosures in 2027.

In Figure below, the Main Sustainability reporting standards described in the Chapter features are summarized with a benchmark approach (Table 2.17).

Table 2.17 Main sustainability reporting standards features. Source: Author Elaboration

Feature	California Climate Disclosure Bills	CSRD/ESRS	IFRS S1 and S2	GRI Standards
Countries/ regions	California, USA	European union	Global	Global
Subjected entities	Companies with significant operations in California or over $1 billion in annual revenue	Large companies and listed entities in the EU	All companies with sustainability-related disclosures	Any organization regardless of size or sector
Type of materiality assessment	Specific to climate-related risks and GHG emissions	Broad sustainability double materiality assessment	General and climate-specific materiality assessment	General sustainability materiality assessment
GHG disclosure	Mandatory disclosure of scope 1, 2 emissions (scope 3 emissions starting in 2027)	Mandatory disclosure of environmental impacts	Mandatory disclosure of climate-related impacts (S2)	Voluntary disclosure of environmental impacts
Scope of reporting	Climate-related financial risks and GHG emissions	Comprehensive sustainability reporting (financial, environmental, social)	Financial, climate-risk and sustainability-related disclosures	Comprehensive sustainability reporting
Assurance requirement	Not explicitly required (but encouraged)	Mandatory assurance of reported data	Not explicitly required (but encouraged)	Not explicitly required (but encouraged)
Standardization	Specific to California regulations	Utilizes European sustainability reporting standards (ESRS)	Utilizes IFRS standards	Utilizes GRI'S universal standards

References

Aon. (2022). *Assessing your company's climate maturity: A three-step strategy to address new climate regulations and disclosure standards.* https://humancapital.aon.com

Cheema-Fox, A., Czasonis, M., Kontu, P., & Serafeim, G. (2023). *Corporate alignment with the EU taxonomy for sustainable activities: First evidence from financial accounting data.* State Street Associates.

Deloitte. (2025). *Adoption of IFRS sustainability disclosure standards by jurisdictions.*

EFRAG & IFRS. (2024). *ESRS-ISSB standards. Interoperability guidance.* https://www.efrag.org/Assets/Download?assetUrl=/sites/webpublishing/SiteAssets/ESRS-ISSB+Standards+Interoperability+Guidance.pdf

European Commission. (2017). *European pillar of social rights.* Publications Office of the European Union. https://op.europa.eu/webpub/empl/european-pillar-of-social-rights/en/

European Parliament and Council of the European Union. (2014). Directive 2014/95/EU of the European Parliament and of the Council of 22 October 2014 amending Directive 2013/34/EU as regards disclosure of non-financial and diversity information by certain large undertakings and groups. *Official Journal of the European Union, L, 330*, 1–9.

European Parliament and Council of the European Union. (2019). Regulation (EU) 2019/2088 on sustainability-related disclosures in the financial services sector. *Official Journal of the European Union, L, 317*, 1–16.

European Parliament and Council of the European Union. (2020). Regulation (EU) 2020/852 on the establishment of a framework to facilitate sustainable investment, and amending Regulation (EU) 2019/2088. *Official Journal of the European Union, L, 198*, 13–43.

European Parliament and Council of the European Union. (2022). Directive (EU) 2022/2464 of the European Parliament and of the Council of 14 December 2022 amending Regulation (EU) No 537/2014, Directive 2004/109/EC, Directive 2006/43/EC and Directive 2013/34/EU, as regards corporate sustainability reporting. *Official Journal of the European Union, L, 322*, 15–80.

Global Reporting Initiative. (2021). *A short introduction to the GRI standards.* https://www.globalreporting.org

Global Reporting Initiative. (2024). *CSRD essentials.* https://www.globalreporting.org/media/nchpzct5/gri-csrd-essentials.pdf

Goldman Sachs. (2025, January 30). *Equity research.*

International Energy Agency. (2016). *Water-energy Nexus: World Energy Outlook special report.* https://www.iea.org/reports/water-energy-nexus

KPMG. (2024). *The move to mandatory reporting - Survey of sustainability reporting 2024.* KPMG International.

Platform on Sustainable Finance. (2022). *Final report on minimum safeguards.* European Commission. https://finance.ec.europa.eu/system/files/2022-10/221011-sustainable-finance-platform-finance-report-minimum-safeguards_en.pdf

PwC. (2025). *VALUE Plc—Example sustainability disclosures under IFRS sustainability disclosure standards.*

Sautner, Z., Yu, J., Zhong, R., & Zhou, X. (2025). The EU taxonomy and the syndicated loan market. *Journal of Financial Services Research.*

An Introduction to Green Finance and the Linked Consulting Services

3

Green Finance is a subset of Sustainable Finance and it is one of the terms used to label activities that imply the interaction between environment, finance, and investments. As a matter of fact, the most appropriate term is Sustainable Finance if Social and Governance topics are also involved.[1] Otherwise, if the only dimension involved is Environment, the most appropriate term is Green Finance.

Addressing climate change and broader sustainability challenges is essential for safeguarding human existence. However, steering economic development onto a sustainable trajectory requires the active engagement and alignment of the financial system with the objectives of the Paris Agreement and the 2030 Agenda for Sustainable Development. Article 2.1(c) of the Paris Agreement explicitly calls for making financial flows consistent with pathways toward low greenhouse gas emissions and climate-resilient development.[2] Similarly, the Sustainable Development Goals (SDGs), adopted by the United Nations in 2015, provide a comprehensive framework for governments, businesses, and civil society to tackle these interconnected challenges (See Fig. 3.1).

Green and sustainable finance instruments play a pivotal role in this transition. To ensure transparency and accountability, these financial products must clearly articulate their intended contribution to the SDGs. According to the Sustainability Bond Guidelines (SBG) developed by the International Capital Market Association (ICMA),[3] issuers are expected to disclose the expected SDG impacts in pre-issuance

[1] Thompson S. (2017) Green and Sustainable Finance, Principles and Practice. Kogan Page, London

[2] United Nations Framework Convention on Climate Change. Paris Agreement. UNFCCC, 12 Dec. 2015

[3] International Capital Market Association. (2021). Sustainability Bond Guidelines.

M. Gussoni, *Integrating Strategic Sustainability Practices in Business*, CSR, Sustainability, Ethics & Governance,
https://doi.org/10.1007/978-3-032-09617-3_3

Fig. 3.1 The UN sustainable development goals (SDGs)—Source: www.un.org

documentation, aligning their frameworks with the core principles of transparency, use of proceeds, and impact reporting.

From a strategic point of view, it is interesting to observe that as illustrated in Fig. 3.2, the EU's progress toward the SDGs is uneven, with varying degrees of advancement across different goals. While significant steps forward have been moved toward the improvement of work conditions and economic growth (SDG 8) and toward inequalities reduction (SDG 10); we cannot say the same for life and land (SDG 15), Good health and well-being (SDG 3), Affordable clean energy (SDG 7), and clean water and sanitation (SDG 6). This means that we can expect increasing investments in these last topics in the next years and an increasing attention by the European investing landscape and governments.

Bonds are debt instruments issued by government or a company to borrow funds from the individual or corporate investors for a specific duration. They are similar to loans where the principal sum (the issuance value) is required to develop a specific project and the issuer offers periodic interest or coupon to the bond holders at the end of the term or maturity period.

Benefits of thematic issuances regard both the issuers and the investors in different ways.

The issuance of a labelled bond, grants to the issuer the access to additional capital sources from dedicated sustainable financing and a diversified investor base; reputational benefits and a stronger investor demand.

For investors, thematic labels assure the integration of climate risks into their project planning which may lead to a reduction in risk and a reduced burden of due diligence; increasing transparency and accountability on the use and management of proceeds and an increase levels of engagement with management, particularly on sustainability issues.

Benefits of thematic issuances are also systemic, stimulating positive market reaction and improvements in financial performance fostering green and social

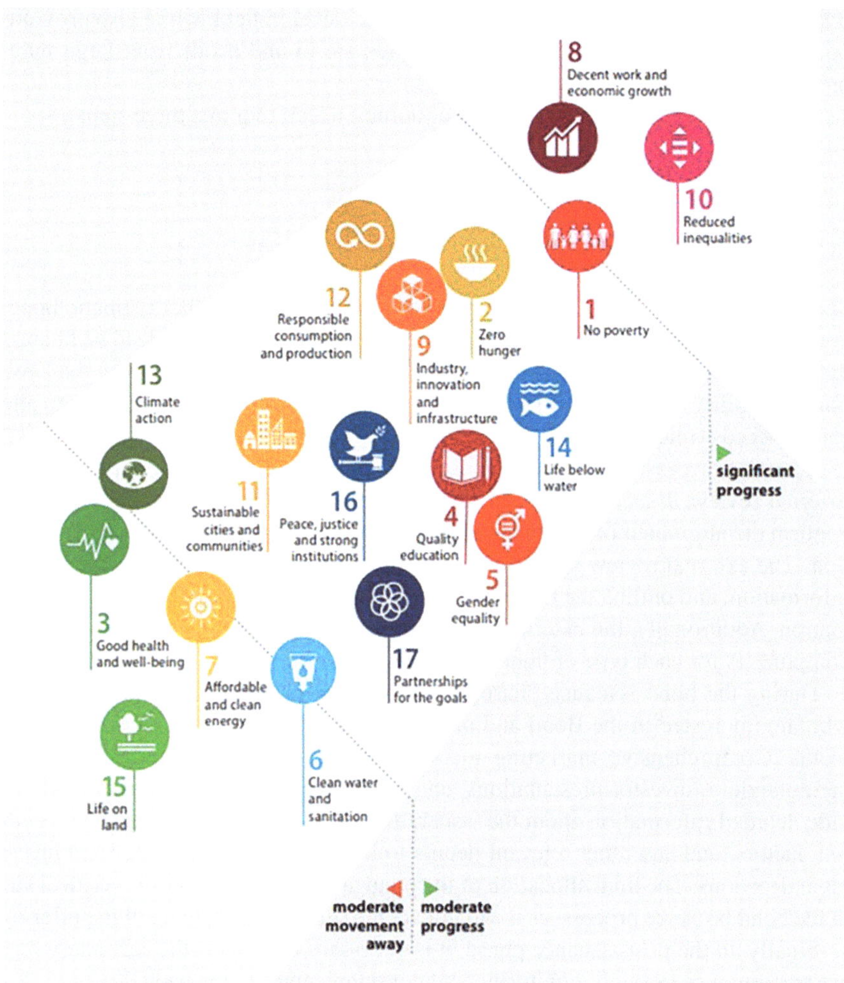

Fig. 3.2 Overview of EU progress toward the SDGs over the past 5 years, 2024 (Data mainly refer to 2017–2022 or 2018–2023). Source: Eurostat (2024) Sustainable development in the European Union Monitoring report on progress toward the SDGs in an EU context. 2024 edition. Pg. 11

innovation and helping transition toward low-carbon, resource-efficient and inclusive economies. This could also develop opportunities for Public–Private partnerships (PPPs).[4] PPPs can be used in a variety of sectors, such as infrastructure (e.g., roads, bridges, hospitals), utilities (e.g., water, energy), and social services (e.g., education, healthcare). They can help governments address infrastructure needs,

[4] A public–private partnership (PPP) is a collaborative arrangement between the public sector (government) and the private sector (businesses or non-profit organizations) to deliver a project or service. In a PPP, the partners share the risks, responsibilities, and rewards of the project or service.

improve public services, and promote economic development, while also providing opportunities for private companies to participate in and profit from large public projects.

The issuance of a thematic Bond is a journey which requires three steps:

- Pre-issuance.
- Issuance.
- Post-issuance.

Speaking of Green Bonds, but the process is similar also for other thematic bonds, during the pre-issuance phase, it is necessary to select projects eligible to be financed through a green bond (selection of projects); define how the projects meet the Green bond eligibility criteria (use of proceeds) and set up accounts and processes to allocate proceeds (management of proceeds).

It is also necessary to establish the reporting process and get a pre-issuance external review. Issuers should, in fact, engage an external review provider(s) to confirm the alignment of the financial instrument with its components and information. The external review should disclose the scope of the review, provide general information, and outline the reporting commitments that form the basis of the information. Additionally, the external review should include an assessment of the core components for each type of financial instrument.

During the bond issuance phase, it is necessary to build the book of investors who are interested in the Bond and prepare marketing material and investor documents. Comprehensive marketing materials should include preliminary offering memorandum, investor presentations, and fact sheets. These materials should provide detailed information about the issuer, the bond's features, the use of proceeds, risk factors, and any other relevant details to help investors make informed investment decisions. The final allocation of the bond among the investors is a critical step in the bond issuance process, as it can impact the success and pricing of the offering.

Finally, in the post-issuance phase, it is necessary to support the secondary market performance through monitoring and communication to the capital market. It is also necessary to allocate the proceeds to the projects and track allocation over time publishing impact reports. The issuer of labelled financial instruments must annually present a report. It is recommended that the reports be validated by the independent external reviewer.

Transparency is of paramount importance when communicating the expected and/or achieved impact of the thematic financial instrument's projects. The ICMA Principles and the Climate Bonds Standard call for issuers to include both qualitative performance indicators and, where feasible, quantitative indicators to showcase the anticipated impact of the financed projects.[5]

[5] Climate Bonds Initiative, Latinex, and IDB Invest. (2024). Guidelines for the issuance of thematic labelled financial instruments.

To establish a harmonized framework for the publication of impact reports on green projects, the ICMA Green Bond Principles Impact Reporting Working Group has developed voluntary guidelines.[6] These guidelines provide recommendations for impact reporting, particularly for projects related to energy efficiency, renewable energy, water and wastewater management, and waste management.

Case Study

Labelled Bond Issuance—The Case of Société des Grands Projets

Société des grands projets is the industrial and commercial public agency, wholly owned by the French State, dedicated to the design, construction, and financing of the Grand Paris Express.

The Grand Paris Express is the largest transit and development project in Europe. It is also the first bond issuer to adopt a 100% green Euro medium-term note (EMTN)[7] programme. The Project aims to realize 200 km of fully electric power, high passenger volume metro lines in the ile-de-France Region through the issuance of green bonds for a value of about 30 bln Euros.

The framework is aligned with the 2021 version of the Green Bond Principles and the June 2022 Appendix published by ICMA, therefore for each green bond issued, the Societe des Grands Projects ensures it complies with the following four pillars:

1) Justification of use of proceeds.
2) Definition of processes for projects evaluation and selection.
3) Detailed description of management of proceeds procedure.
4) Reporting commitments.

Societè des Grand Projects has also chosen to align their framework to the EU Taxonomy for the activity covered by the section "Infrastructure enabling low-carbon low transport and public transport" contributing to the objectives of climate change mitigation and climate change adaptation.

Finally, Sustaianalytics provided verification and certification detailing how the project meets Climate Bonds Standards.[8]

All these steps are aimed at granting more transparency and gain credibility among investors.

[6] Green Bond Principles, ICMA (2022) Handbook—Harmonized Framework for Impact Reporting. Available at: https://www.icmagroup.org/

[7] A Euro medium-term note is a flexible debt instrument typically issued by corporations, financial institutions, or governments in the European markets. Unlike traditional bonds, EMTNs are issued under a program that allows multiple issues of securities over time.

3.1 The Sustainable Bond Market

The debt capital markets play a vital role in allocating capital to finance solutions to environmental, social, and transition challenges. They can help investors to diversify and manage risk in their portfolio and to adopt responsible investments approaches.

Sustainable bonds refer to a broad category of bonds where the proceeds are used to finance a mix of green and social projects. They include green bonds, social bonds, sustainability bonds, and sustainability-linked bond as subcategories. It is also worth to mention Climate Bond which is oriented to finance projects aiming at climate change mitigation and adaptation with, thus, a specific focus on reducing greenhouse gas emissions or increasing climate resilience.

According to the latest Global Sustainable Investing Alliance (GSIA) Report,[9] global sustainable investing assets,[10] in 2022, reach the value of 30.3 trillion USD dollars, the 24.4% of the global Asset Under Management (AuM) whose value is 124.5 trillion USD Dollars.

In 2024, the Sustainable Bond Issuance[11] reaches a value of 854 USD bl, from 788.4 USD bl in 2023 (+8.3%) (Source: International Capital Market Association (ICMA) Sustainable Bond Market Data). 46.3% of the total value is due to Sustainable Bond European issuances.

The key characteristics of sustainable bonds are:

1. **Use of Proceeds**: The funds raised from the bond issuance must be used to finance or refinance eligible green or social projects or a combination of both. These can include renewable energy, energy efficiency, clean transportation, affordable housing, access to essential services, and other sustainable initiatives.
2. **Project Evaluation and Selection:** The issuer must have a clear process for evaluating and selecting the eligible projects that align with the issuer's sustainability objectives and principles.
3. **Management of Proceeds**: The issuer must have a system in place to track and monitor the allocation of the bond proceeds to the eligible projects.

---

[8] The Climate Bonds Standard and Certification Scheme is a voluntary labelling scheme for investments—and now entities—that addresses the challenge of climate change and is consistent with the goals of the Paris Agreement. The Scheme was launched in 2012 by the Climate Bonds Initiative (Climate Bonds), a not-for-profit organisation working to mobilize global capital for climate action.

[9] Global Sustainable Investing Alliance (2022) Global Sustainable Investment Review 2022

[10] The sustainable bond asset value refers to the total market value of all outstanding sustainable bonds. This includes green bonds, social bonds, and sustainability-linked bonds. The asset value represents the current market price of these bonds.

[11] The sustainable bond issuance value refers to the total amount of new sustainable bonds that have been issued in a given period, typically a year.

4. **Reporting:** The issuer is required to provide regular reporting on the use of proceeds and the environmental or social impact of the funded projects.

Sustainable bonds can take different forms, such as Green Bonds, Climate Bonds, Social Bonds, Sustainability Bonds, and Sustainability-linked Bond depending on the specific focus of the projects being financed. These bonds are designed to provide investors with the opportunity to allocate capital toward sustainable development while also offering issuers a way to raise funds for their sustainability initiatives.

As reported in Fig. 3.3, Green bond issuance represents 57.51% of the total value, followed by Sustainability Bond (23.2%), Social Bond (15.49%), and Sustainability-linked Bond (3.8%) (Source: ICMA Sustainable Bond Market Data).

Green bonds are therefore the most spread debt instruments. In Green Bonds, the proceeds are exclusively used to fund projects with clear environmental benefits. These projects are typically aligned with the four core components of the Green Bond Principles (GBP) (ICMA, Green Bond Principles—Voluntary Process Guidelines for Issuing Green Bonds, 2022) established by the International Capital Market Association (ICMA) (ICMA, Guidance Handbook, April 2020) to grant transparency and disclosure.

There are four types of Green Bond:

- **Standard Green Use of Proceeds Bond**: Debt instruments in which the repayment and payment of interest to investors depend exclusively on the solvency and repayment capacity of the issuer, without any constraint on specific cash flows or assets.
- **Green Revenue Bond**: Debt instrument in which the repayment and payment of interest to investors depend exclusively on the cash flows generated by specific revenues, fees, or taxes, without any recourse to the issuer. This type of bond instrument allows to finance green projects without directly burdening the issuer's balance sheet.
- **Green Project Bond:** A type of bond instrument where the proceeds are used to finance a single green project or multiple green projects. In this case, the investors have direct exposure to the risk of the underlying green project(s), rather than the general creditworthiness of the bond issuer.

Fig. 3.3 Sustainable bond issuances per category 2024. Source: Author elaboration on the basis of ICMA Data

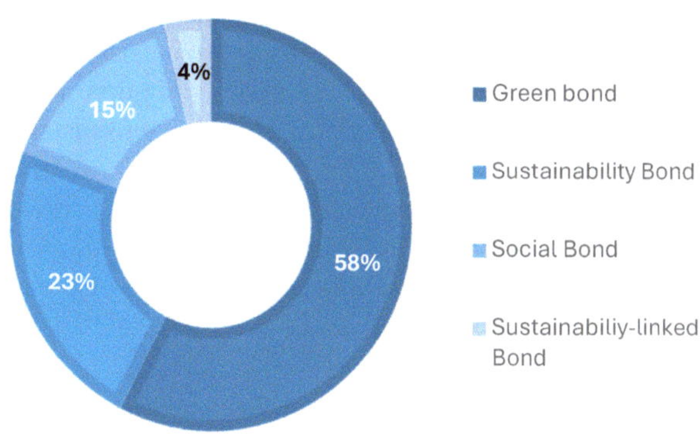

- **Secured Green Bond:** A type of bond where the net proceeds are exclusively used to finance or refinance green projects, and the bond is secured by specific collateral or assets.

All the above have to be aligned with the GBP, whose key four components are:

1. Use of Proceeds;
2. Process for Project Evaluation and Selection;
3. Management of Proceeds;
4. Reporting.

The eligible Green Projects categories include, but are not limited to: renewable energy; energy efficiency; pollution prevention and control; environmentally sustainable management of living natural resources and land use; terrestrial and aquatic biodiversity conservation; clean transportation; sustainable water and wastewater management; climate change adaptation; circular economy adapted products, production technologies and processes and/or certified eco-efficient products; green buildings (ICMA, Green Bond Principles—Voluntary Process Guidelines for Issuing Green Bonds, 2022).

Focus

Green Financing: The Case of ENEL Group

Enel Group, the Italian multinational energy company, has been a pioneer in the green bond market. As one of the world's largest private renewable energy operators, Enel has utilized green bonds extensively to finance its transition toward a more sustainable energy portfolio.

Since 2017, Enel has issued three Green Bonds, for a total of 3.5 billion euros. In November 2018, Enel prepared and published a new Green Bond Framework, updated with the presentation of the 2019–2021 Strategic Plan, with the aim of promoting the Group's transparency and commitment in relation to green bonds[12] (See: Enel Green Bond Report 2023).

Following this updated framework, the eligible green projects of the Enel Group have been identified, in accordance with the "Green Bond Principles" published by the International Capital Market Association (ICMA). Specifically, among the projects identified, for example, projects for the development of generation plants from renewable sources, construction and management of transmission and distribution networks, smart metering systems and development of projects relating to sustainable mobility and demand response.

Enel's Green Bond Framework has been verified by Vigeo Eiris. In 2017, Enel established a Green Bond Committee to oversee the implementation of the Green Bond Framework and the process of allocating green bond proceeds.

[12] https://www.enel.com/it/investitori/investimenti/finanza-sostenibile/green-bond

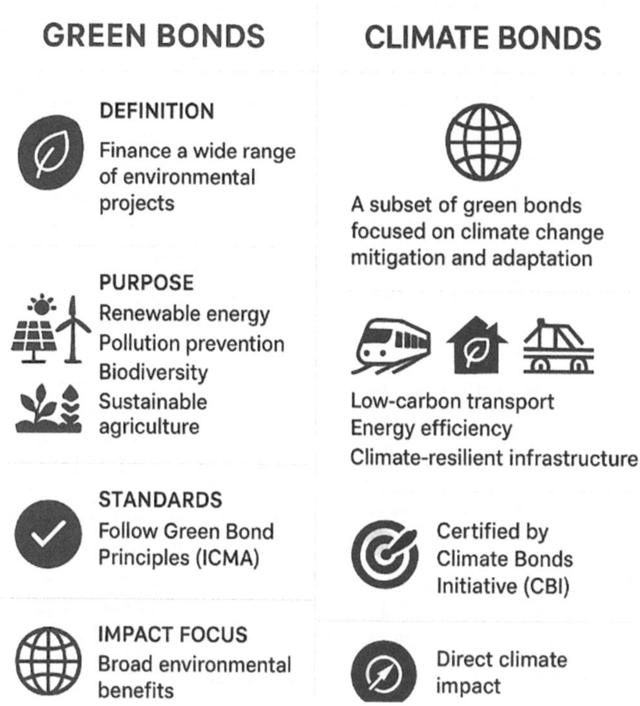

Fig. 3.4 Differences between green and climate bonds. Source: Author Elaboration

A subset of Green Bonds is represented by the **Climate Bonds** which are oriented to finance projects aiming at climate change mitigation and adaptation with, thus, a specific focus on reducing greenhouse gas emissions or increasing climate resilience.

The purpose of Climate Bonds is, for instance, to finance projects aimed at developing low-carbon transport, increase energy efficiency, build climate-resilient infrastructure, or deploy carbon capture technologies. They must comply with the Climate Bonds Standard (CBS) and they are certified by the Climate Bond Initiative (CBI) (Fig. 3.4).

3.2 Defining a Sustainable Bond Framework

A sustainable bond framework is a set of guidelines and principles that a company or organization follows when issuing bonds to raise funds for projects or activities that have a positive environmental or social impact.

The framework typically includes:

1. An introduction section, including statements on issuer's environmental objectives and environmental strategy.

2. Use of Proceeds: It must outline how the funds raised from the bond will be used, such as for renewable energy projects, green buildings, or social initiatives like affordable housing or healthcare access.
3. Project Evaluation and Selection: It describes the process the issuer uses to evaluate and select eligible projects that align with the framework's sustainability criteria.
4. Management of Proceeds: It explains how the bond proceeds will be managed and tracked to ensure they are allocated to the intended sustainable projects.
5. Reporting: It commits the issuer to regularly report on the use of proceeds and the environmental or social impacts of the funded projects.
6. External Review.

The sustainable bond framework helps investors understanding the sustainability credentials of the bond and provides transparency on how the funds will be used. It also demonstrates the issuer's commitment to sustainability and responsible financing.

By following a well-defined framework, issuers can access a growing pool of investors who are specifically looking for sustainable investment opportunities, which can help them raise funds at more favorable terms.

The sustainable bond framework is typically defined and developed by the issuing organization itself, often with the support of external consultants or advisors. The service of developing social and sustainable finance frameworks and sustainable finance strategies is common among big consulting companies (e.g., BCG, KPMG, Deloitte, etc.)

The International Capital Market Association (ICMA), however, in its Green Bond Principles and Social Bond Principles, states that the issuer is responsible for defining and disclosing the sustainable bond framework.

The following sub-chapters describe the structure of a sustainable bond framework in order to help issuers and consultants in the development phase.

3.2.1 Introduction Section, Background of the Issuer

A Sustainable Bond Framework usually starts with an introduction on the issuer's environmental objectives and strategy. The introduction includes background general information about the entity issuing a bond (geographic presence, number of employees, financial performance, and credit rating); the description of the policies and commitments to fight the climate change (e.g., emission targets, SDGs targeted, environmental and social risks assessment) and why the entity is issuing a debt.

Often the description of the context for issuing a green, social, or sustainability bond refers to the issuer need to meet sustainable finance targets, realizing a pipeline of projects.

An important introductory section regards the description of the impacts of the issuer to the local economy, like corporate social initiatives, capacity building efforts, etc.

Finally, it is important to describe the alignment to national regulation and targets (e.g., support for national climate change mitigation and adaptation goals; compliance with relevant environmental, social, and governance regulations).

By providing this comprehensive introduction, the issuer demonstrates its commitment to sustainability, the strategic importance of the sustainable bond, and the potential positive impacts it aims to achieve. This sets the stage for the subsequent sections of the framework, which outline the specific use of proceeds, project selection criteria, and reporting commitments.

Incorporating these elements in the introduction helps to build investor confidence and aligns the sustainable bond with the issuer's overall sustainability vision and performance.

3.2.2 Use of Proceeds

This section has to define the green categories under which eligible projects fall, i.e. broad categories of eligible projects or assets that the bond proceeds can be allocated toward, such as renewable energy, green buildings, clean transportation, affordable housing, or access to essential services. These categories are often aligned with industry standards and taxonomies, such as the Green Bond Principles, Social Bond Principles, or the EU Taxonomy for Sustainable Activities.

By clearly defining the eligible project categories, the selection process, and the reporting commitments, the use of Proceeds section provides investors with the necessary information to assess the sustainability credentials of the bond and the intended use of the funds. It is important to underline that bond issuers face high reputational risks if bonds do not deliver the intended environmental benefits. Investors may seek penalties if the issuer breaks agreed green clauses.

In Fig. 3.5, it is reported a well-structured example of "Use of Proceed" section in a Financial group's Sustainable Bond Framework. The eligibility criteria under this Framework are designed to align with the Substantial Contribution Criteria outlined in the EU Taxonomy Climate Delegated Act from June 2021. In addition to meeting the Substantial Contribution criteria, the Eligible Green Assets also aim to adhere to the Do No Significant Harm criteria and Minimum Social Safeguards, where practically feasible. The Eligible Green Projects have also been mapped to the relevant UN Sustainable Development Goals (SDGs) and the corresponding NACE codes included in the EU Taxonomy Climate Delegated Act.

3.2.3 Project Evaluation and Selection

The section, dedicated to the description of the process of project evaluation and selection, has to describe the internal set-up and specific governance mechanisms for the selection of projects and assets.

This process involves several steps and layers of review:

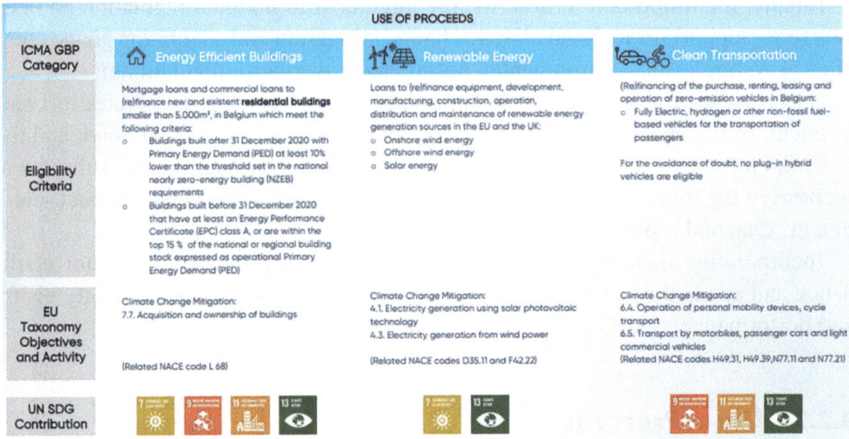

Fig. 3.5 Example of "Use of Proceed" Section—Source: KBC (2023) "KBC Green Bond Framework" available at: https://wcmassets.kbc.be/content/dam/kbccom/doc/investor-relations/7-Debt-issuance/Green-Bonds/20240109-kbc-gb-investor-presentation.pdf.cdn.res/last-modified/1704792133283/20240109-kbc-gb-investor-presentation.pdf, pg. 19; www.kbc.com

1. **Identification of Potential Projects**: The relevant business units and departments within the organization identify potential green projects and assets that align with the eligibility criteria defined in the Framework. Based on typical organizational structures and green financing practices, the relevant business units and departments responsible for the initial identification of potential Eligible Green Projects are usually: Sustainability/ESG Department; Project Development/Engineering Department; Operations Department; Facilities Management; Research and Development (R&D); Corporate Strategy.
2. **Evaluation by the Sustainability/ESG Team:** The Sustainability or ESG team reviews the identified projects and assets to assess their adherence to the Substantial Contribution, Do No Significant Harm, and Minimum Social Safeguards criteria of the EU Taxonomy. This evaluation also considers the mapping to relevant UN SDGs.
3. **Review by the Green Finance Committee:** A dedicated Green Finance Committee, composed of representatives from Finance, Sustainability, Risk, and other relevant functions, reviews the evaluated projects and assets. The Committee provides final approval on the selection of Eligible Green Projects and Assets.
4. **Ongoing Monitoring**: The selected eligible Green Projects and Assets are subject to ongoing monitoring to ensure continued alignment with the Framework's criteria. The Sustainability/ESG team and Green Finance Committee regularly review the portfolio.

3.2.4 Management of Proceeds

In the section "Management of Proceeds," it is described the internal function responsible for the proceed management and it is explained how the proceeds will be segregated, either in a dedicated bank account or through an internal tracking system, to ensure they are allocated only to eligible initiatives.

It is also important to clarify how any unallocated proceeds will be temporarily held and invested, until they are fully allocated to Eligible Green Projects and Assets, specifying the types of temporary investments that may be used, such as cash, cash equivalents, or other liquid instruments.

It is worth to underline that it is necessary to ensure that the temporary investment of unallocated proceeds will not be used for purposes that conflict with the environmental and social objectives of the green financing.

Another important point is the description of the allocation reporting process with the commitment to providing regular (e.g., annual) allocation, reporting to investors and other stakeholders. It is important to outline the information that will be included in the allocation reporting, such as the total amount of proceeds allocated, the list of Eligible Green Projects and Assets funded, the amount of unallocated proceeds (if any) and the types of impact metrics and key performance indicators (KPIs) that will be used to measure and report on the environmental and social benefits of the Eligible Green Projects and Assets. Finally, the process for verifying the allocation of proceeds, which may involve internal or external audits or reviews, must be explained.

3.2.5 Reporting

This section describes the reporting process to be followed in the post-issuance phase.

Reporting should occur, at least, once a year and until full allocation of proceeds. Reports should be made available to the bond's investors (e.g., publication on the issuer's website) providing information in a clear and easy-to-find way on both the allocation of proceeds (Use of Proceeds or UoP) and their expected environmental impacts. Having a dedicated green bond webpage, with all the relevant material related to the issuer's green bond issuance, including clear descriptions and links to all documents, is highly advised (Climate Bond Initiative, 2021).[13]

According to the ICMA (2022) *Harmonised Framework for Impact Reporting,*[14] the report should indicate the total signed amount and the amount of green bond proceeds allocated to eligible disbursements, as well as provide additional details, like the year of signing or project financing stage. Issuers are encouraged to have a

[13] Climate Bond Initiative (2021). Post-Issuance reporting in the green bond market

[14] In March 2015, a working group of four Multilateral Development Banks (AfDB, EIB, IFC, and World Bank) developed and published the first Harmonised Framework for Impact Reporting that was later revised and republished by 11 International Financial Institutions (IFIs). The framework outlined core principles and recommendations for impact reporting in order to provide issuers with reference and guidance for the development of their own reporting and provided core indicators and reporting templates for Energy Efficiency and Renewable Energy Projects.

formal internal process for allocating proceeds and to report on this allocation, potentially providing a list of projects or reporting at a portfolio level.

Focus

Danske Bank Green Bond Report 2024[15]

A good example of Green Bond Report is that published by the Danske Bank in 2024.

The report starts with a letter of acknowledgment to green bond investors by the Bank's head of sustainable finance; after that, it presents some key quantitative data about the allocation, by sector and by country, of the group's green loans and bonds proceeds. Data on the total volume of green loans and bonds issued are also provided.

As shown in the Fig. 3.6, a particular attention has been dedicated to the estimate of the environmental impacts derived from the use of proceeds by sector:

- The annual renewable energy produced;
- The annual energy saved and transmitted;
- The waste treated;
- The energy storage capacity;
- The avoided GHG emissions;
- The impacts on SDGs.

The Report provides also the description of the methodology used for these estimations and a summary of the Bank's green finance framework.

The Independent Auditor's Assurance Report on the green loan allocation reporting closes the document.

Green loan categories — Danske Bank

	Use of proceeds (EUR millions)	Annual renewable energy produced (MWh)	Annual energy savings (MWh)	Waste treated annually (t)	Energy storage capacity (MW)	Annual energy transmitted (MWh)	Annual GHG emissions avoided (tCO2e)	SDGs
Clean transportation	1,107.0						39,015	
Renewable energy	1,932.3	5,028,282			1,121	711,599,287	836,774	
· Hydropower	328.9	1,091,541					208,484	
· Solar and wind	638.8	1,304,308					249,123	
· Bioenergy	513.2	2,632,433					379,167	
· Energy transmission	302.1					711,599,287		
· Energy storage	141.4				1,121			
· Other	7.9							
Green buildings	3,422.9		24,068				3,240	
Pollution prevention and control	126.1			561,443				
Environmentally sustainable management of living natural resources and land use	30.4						73,500	
Total	**6,618.7**	**5,028,282**	**24,068**	**561,443**	**1,121**	**711,599,287**	**952,529**	
Assessed tonnes CO_2e savings per EUR millions							**143.9**	

Fig. 3.6 Quantitative datatable example from Danske Bank Green Bond Report 2024, pg. 6

[15] Available at https://danskebank.com

Impact reporting should therefore illustrate the expected environmental impacts or outcomes of the funded projects, based on ex ante estimates. Issuers are encouraged to report on sector-specific core KPIs, while being transparent about the methodologies and assumptions used, as comparing impacts across projects and portfolios can be challenging. Issuers should also be transparent about projects with partial eligibility and may attribute results to different components based on their relative share of financing.

The report should include qualitative information to provide context on the projects' impacts and management of risks. Issuers may facilitate data collection and transfer by using standardized reporting templates or impact reporting databases.

3.2.6 External Review

The External review is an independent assessment of green credentials of a bond provided to the issuer by an external auditor (reviewer).

This process is crucial for ensuring that the bond meets specific environmental standards and criteria, thereby enhancing its credibility and attractiveness to environmentally conscious investors. The external review methodology could be: (i) a Second Party Opinion; (ii) Rating and Scoring, and (iii) Assurance. The last one brings to a certification.

A **Second Party Opinion (SPO)** is an independent evaluation provided by an external expert or organization that assesses the environmental credentials of a green bond. It assesses the environmental credentials of the green bond before issuance.

Rating and Scoring involve the assessment of a green bond's environmental performance by specialized rating agencies. It evaluates the green bond's environmental performance against established benchmarks.

Assurance is a process where an independent auditor verifies the issuer's environmental claims regarding the green bond. It is typically used in impact reporting

The external review in the Impact Reporting typically regards the Assurance Process and verifies the environmental impact claims made by the issuer. The Process of assurance involves some key steps:

1. **Selection of Assurance Provider**
   The issuer selects an independent assurance provider with expertise in environmental standards and green finance. This provider could be an auditing firm or a specialized environmental consultancy.
2. **Planning the Assurance Engagement**:
   - **Scope Definition**: The assurance provider and the issuer agree on the scope of the assurance engagement, including which projects and metrics will be reviewed.
   - **Criteria and Standards**: The assurance provider identifies the criteria and standards against which the impact report will be assessed (e.g., Green Bond Principles, Climate Bonds Standard).

3. **Conducting the Assurance**:
   - **Document Review**: The assurance provider reviews the documentation and data provided by the issuer to verify the reported environmental impacts.
   - **Site Visits**: If necessary, the assurance provider conducts site visits to inspect the projects and verify that they are being implemented as described.
   - **Interviews**: The assurance provider may interview key personnel involved in the projects to gather additional information and clarify any uncertainties.
4. **Verification and Analysis**:
   - The assurance provider analyzes the data and documentation to verify the accuracy and reliability of the reported environmental impacts.
   - **Cross-Checking**: The provider cross-checks the reported data with independent sources or benchmarks to ensure consistency and credibility.
5. **Drafting the Assurance Statement**:
   - **Findings and Conclusions**: The assurance provider drafts a statement summarizing their findings and conclusions. This includes an assessment of whether the reported impacts are accurate and whether the projects meet the specified environmental criteria.
   - **Recommendations**: The assurance provider may include recommendations for improving the reporting process or addressing any identified issues.
6. **Issuance of the Assurance Statement**:
   - The final assurance statement is issued to the issuer, who includes it in the annual impact report. This statement enhances the credibility of the report and provides assurance to investors regarding the environmental integrity of the green bond.
7. **Publication and Communication**:
   - The issuer publishes the annual impact report, including the assurance statement, and communicates the results to stakeholders and investors. This transparency helps build trust and confidence.

References

Climate Bond Initiative. (2021). *Post-Issuance reporting in the green bond market.*

Climate Bonds Initiative, Latinex, and IDB Invest. (2024). *Guidelines for the issuance of thematic labelled financial instruments.*

Global Sustainable Investing Alliance. (2022). *Global sustainable investment review 2022.*

Green Bond Principles, ICMA. (2022). *Handbook—Harmonized framework for impact reporting.* https://www.icmagroup.org/

International Capital Market Association. (2021). *Sustainability bond guidelines.* ICMA Group.

Sustainability Reporting Consulting in Practice

4

4.1 How to Perform a Double Materiality Assessment

The concept of double materiality is a crucial requirement prescribed by the Corporate Sustainability Reporting Directive (CSRD) for the identification of the content to be included in extra-financial reporting.

Materiality analysis is a multi-purpose tool for prioritizing sustainability issues from the double perspective of companies and stakeholders, meaning that both parties contribute to identifying the present and emerging social and environmental risks and opportunities (Calabrese et al., 2019[1]).

The understanding of the main impacts, risks, and opportunities faced by the company, in the context of the expectations and interests of its various stakeholders, can provide, in fact, valuable insights. These insights can inform the company's future strategic positioning and guide the selection of potential strategic choices. By thoroughly analyzing the key factors that influence the company's performance and reputation from the perspective of its stakeholders, such as customers, employees, investors, regulators, and the local community, the company can gain a more comprehensive understanding of the landscape in which it operates. This understanding can help the company anticipate and prepare for potential challenges, as well as identify opportunities to create value and strengthen its competitive position.

The strategic choices made by the company can then be informed by this stakeholder-centric analysis, ensuring that the company's long-term direction and decision-making are aligned with the needs and expectations of its key

[1] Calabrese, A., Costa, R., Levialdi Ghiron, N., & Menichini, T. (2019). Materiality analysis in sustainability reporting: A tool for directing corporate sustainability towards emerging economic, environmental and social opportunities. Technological and Economic Development of Economy. Article in press. https://doi.org/10.3846/tede.2019.10550

M. Gussoni, *Integrating Strategic Sustainability Practices in Business*, CSR, Sustainability, Ethics & Governance,
https://doi.org/10.1007/978-3-032-09617-3_4

stakeholders. This alignment can contribute to the company's long-term sustainability, resilience, and success in the market.

The Double Materiality Assessment is the process that requires companies to consider sustainability from two perspectives:

- **Financial Materiality (outside-in):** How sustainability issues impact the company's financial performance.
- **Impact Materiality (inside-out):** How the company's activities impact people and the environment.

The concept of financial materiality focuses on how environmental, social, and governance (ESG) factors can influence the economic value and financial performance of a company.

This perspective is particularly relevant for investors and other financial stakeholders, as it examines how ESG-related risks and opportunities can impact the company's financial health and, ultimately, the returns and value realized by these stakeholders. By considering financial materiality, companies can gain a deeper understanding of how ESG factors, such as climate change, labor practices, or corporate governance, can affect their bottom line, cost of capital, access to financing, and overall market positioning. This understanding can then inform strategic decision-making, risk management, and the allocation of resources to address material ESG-related issues.

Investors and other financial stakeholders can use the insights derived from financial materiality analysis to make more informed investment decisions, assess the long-term viability of a company, and engage with management on ESG-related matters that may impact the company's financial performance and shareholder value. Overall, the financial materiality perspective helps bridge the gap between ESG considerations and the core financial interests of investors and other financial stakeholders, enabling a more holistic and value-driven approach to corporate sustainability and performance.

The concept of Impact Materiality, although not explicitly stated, was already foreseen in NFRD (Directive 2014/95/EU) which focuses on the company's impact on the outside world (*inside-out*). The concept of impact materiality focuses, in fact, on understanding how a company's activities and operations affect a wide range of stakeholders, including communities, employees, customers, and the natural environment. This perspective goes beyond the financial implications and examines the broader social and environmental impacts of the company's actions.

Impact materiality covers a wide range of issues, such as:

- Carbon emissions and climate change impacts;
- Labor practices and working conditions;
- Human rights and social equity;
- Ecological impacts and biodiversity preservation;
- Product safety and customer well-being;
- Community engagement and social development.

By considering impact materiality, companies can gain a more comprehensive understanding of their role and responsibilities within the broader societal and environmental context in which they operate. This understanding can inform strategic decision-making, risk management, and the development of policies and initiatives that aim to mitigate negative impacts and enhance positive contributions to stakeholders.

Addressing impact materiality is crucial for companies to demonstrate their commitment to sustainable and responsible business practices, and to build trust and credibility with a wide range of stakeholders, including regulators, civil society organizations, and the general public (Fig. 4.1).

EFRAG ESRS1 General Requirements Report,[2] and EFRAG's final Materiality assessment implementation guidance published in February 2024,[3] propose a three-step process toward the identification of an entity's material topics:

1. **Understanding the context:** This step involves understanding the context in which the entity operates, including relevant sustainability matters, Impacts, risks, and opportunities (IROs).
2. **Identification of the actual and potential IROs related to sustainability matters:** This step focuses on identifying the actual and potential IROs related to sustainability matters that are relevant to the entity.
3. **Assessment and determination of material IROs related to sustainability matters:** In this final step, the entity assesses and determines which of the identified IROs are material and should be the focus of the sustainability reporting. A sustainability matter can be material from an impact perspective, from a financial perspective, or from both. Most of the material matters also give rise to financial risks and/or opportunities.

Fig. 4.1 Double materiality assessment—Source: ECS 2024 (Euronext Corporate Solutions (2024). The Ultimate guide to double Materiality Assessment)

[2] EFRAG (2022). ESRS1 General Requirements

[3] EFRAG. (2024). Materiality Assessment Implementation Guidance. Retrieved from https://www.efrag.org

The guidance also proposes additional steps that should be included, especially ahead of the actual materiality assessment, such as:

1. Formalization and visualization of the undertaking's business model and value chain.
2. Identification of the company's business ecosystem and its main stakeholders.

Adding these steps is crucial to identify all applicable potentially material topics and thus to apply the principle expressed in ESRS 1 (Fig. 4.2).

With these premises, the steps to be followed for the materiality assessment have been detailed in the following table (Table 4.1):

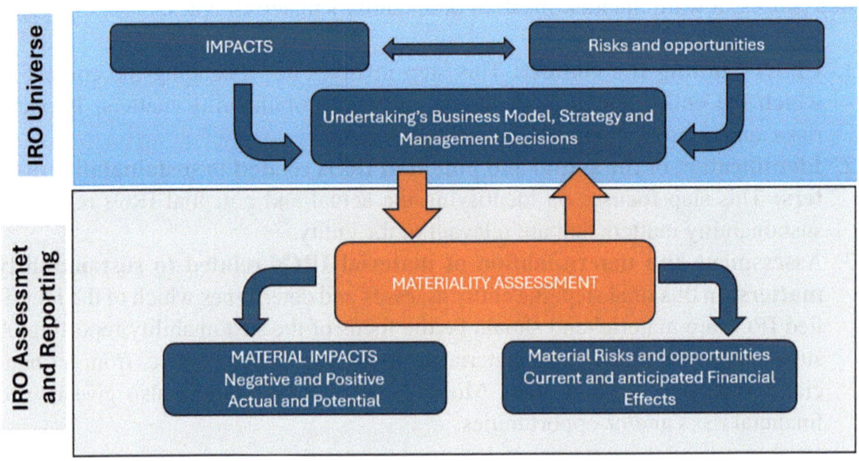

Fig. 4.2 IROs identification and relationship between undertaking's business model and the materiality assessment—Source: Author Elaboration on the bases of EFRAG (2024) (EFRAG. (2024). Materiality Assessment Implementation Guidance. Retrieved from https://www.efrag.org)

Table 4.1 Steps for double materiality assessment. Source: Author Elaborations on the basis of EFRAG Materiality Assessment Implementation Guidance

EFRAG steps	Sub-steps
Understanding the context	1) Analysis of the business model and the value chain (See Chapter 7) 2) Stakeholder mapping (See Sect. 4.2)
Identification of the actual and potential IROs	1) Identification of potentially material topics 2) Stakeholder engagement (See Sect. 4.2)
Assessment and determination of material IROs	1) Impact materiality assessment 2) Financial materiality assessment 3) Collection aggregation and results' analysis from interviews to stakeholders 4) Formalization of the material matrix 5) Identification of relevant indicators for monitoring material IROs

Materiality assessment is therefore the starting point for sustainability reporting under ESRS. It is important to underline that when performing materiality assessment, the undertaking shall at least consider the following list of sustainability matters covered in the topical ESRS[4] and the relative sub-topics (Table 4.2; see further details on topical ESRS in Sect. 2.2 of this book):

Table 4.2 Sustainability matters covered on topical ESRS—Source: DELEGATED REGULATION (EU) 2023/2772

Topical ESRS	Topic	Sub-topics
ESRS E1	Climate change	— Climate change adaptation — Climate change mitigation — Energy
ESRS E2	Pollution	— Pollution of air — Pollution of water — Pollution of soil — Pollution of living organisms and food resources — Substances of concern — Substances of very high concern — Microplastics
ESRS E3	Water and marine resources	— Water — Marine resources
ESRS E4	Biodiversity and ecosystems	—Direct impact drivers of biodiversity loss — Impacts on the state of species —Impacts on the extent and condition of ecosystems —Impacts and dependencies on ecosystem services
ESRS E5	Circular economy	— Resources inflows, including resource use —Resource outflows related to products and services — Waste
ESRS S1	Own workforce	— Working conditions — Equal treatment and opportunities for all — Other work-related rights
ESRS S2	Workers in the value chain	— Working conditions — Equal treatment and opportunities for all — Other work-related rights
ESRS S3	Affected communities	— Communities' economic, social, and cultural rights — Communities' civil and political rights — Rights of Indigenous peoples
ESRS S4	Consumers and end-users	— Information-related impacts for consumers and/or end-users — Personal safety of consumers and/or end-users — Social inclusion of consumers and/or end-users
ESRS G1	Business conduct	— Corporate culture — Protection of whistle-blowers — Animal welfare — Political engagement and lobbying activities — Management of relationships with suppliers including payment practices — Corruption and bribery

[4] DELEGATED REGULATION (EU) 2023/2772, Sustainability matters to be included in the materiality assessment, AR 16

Using this list does not replace the process of identifying material matters. Instead, it serves as a tool to aid the undertaking's materiality assessment.

The undertaking, however, must still consider its unique circumstances when determining material matters. Additionally, the undertaking should, if necessary, develop entity-specific disclosures on significant impacts, risks, and opportunities not covered by ESRS. To support this process, conducting a mapping exercise of peers' ESG topics and relevant research against ESRS requirements can be highly valuable.

Selecting the right material ESG issues is a critical yet complex step in the materiality assessment process. While it may be tempting for organizations to include a broad range of topics, doing so can dilute focus and reduce the strategic clarity of sustainability reporting.

To avoid an overly dispersive approach, companies are encouraged to define a focused shortlist of material topics. One effective method is to conduct sector-specific benchmarking, analyzing the frequency and emphasis of ESG topics disclosed by peer organizations. This data-driven approach helps identify which issues are most commonly considered material within the industry and can serve as a foundation for a company's own prioritization. Moreover, by combining peer analysis with stakeholder engagement and internal strategic alignment, companies can ensure their materiality assessments are both relevant and impactful. According to Dewi et al. (2023),[5] while materiality disclosure significantly influences the quantity of information disclosed, its impact on the overall quality of sustainability reporting depends on how well the information is contextualized and integrated.

Another recommendation is to structure a robust and well-defined materiality assessment process. This involves engaging with key stakeholders, analyzing industry trends, and prioritizing the most significant economic, environmental, and social impacts defining a materiality threshold on the basis of an estimated score. The impact materiality score (See equation below) can be estimated through a survey to stakeholders. In the process of elaborating the questionnaire for stakeholders, for the double materiality assessment, each potentially material topic can be addressed from 4 dimensions as per CSRD requirements: scope, scale, irremediability, and likelihood. Irremediability is part of the evaluation of the severity of the negative impacts. Considering the scope, questions have to be designed to assess the breadth of impact, ranging from local to widespread. Considering scale, questions have to be tailored to gauge the gravity of this impact, whether it be low, moderate, or high. Finally, regarding likelihood, questions have to be formulated to evaluate the likelihood of the impact, ranging from very unlikely to certain.

$$Impact\ Materiality = Scale(score) + Impact(score) * Likelihood(score)$$

[5] Dewi, A. A., Saraswati, E., Rahman, A. F., & Atmini, S. (2023). Materiality, stakeholder engagement disclosure, and corporate governance: Critical elements for the quality of sustainability reporting. Cogent Business & Management, 10(1), 2175437. https://doi.org/10.1080/23311975.2023.2175437

Usually, scores ranges between 1 (very low) and 5 (very high).

For the Financial Materiality assessment, it is recommended to seek input from subject matter experts. The primary stakeholders for this assessment include the CFO, Sustainability Officer, and Risk Officer. The analysis can be conducted either individually or through a workshop involving representatives from all three areas of expertise.

The financial materiality questionnaire is structured as follows:

- The same topics, sub-topics, or sub-sub-topics are used to align impact and financial materiality results.
- Questions are designed to assess the severity and likelihood of potential financial impacts on revenues, costs, cash flow, and access to capital.

The severity of financial impacts is evaluated using monetary thresholds expressed as a percentage of revenues, costs, cash flows, and cost of/access to capital.

These thresholds are proposed by the project team for review and validation or are based on existing risk metrics within the company's current Risk Management framework to ensure alignment with the organization's strategic objectives and risk tolerance.

The Financial Materiality score can be estimated on the basis of the equation below:

$$\text{Financial Materiality} = MAX\left(\text{Score Risk}^{*}\text{Likelihood};\text{Score Opportunity}^{*}\text{Likelihood}\right)$$

Where

- ***Score Risk****: A numerical value representing the potential negative financial impact of a specific ESG issue (e.g., regulatory fines, reputational damage, and operational disruption).*
- ***Score Opportunity****: A numerical value representing the potential positive financial impact (e.g., revenue growth from green products, cost savings from energy efficiency).*
- ***Likelihood****: The estimated probability that the risk or opportunity will materialize.*

Once impact and financial materiality have been assessed, it is necessary to present the results of the activity. Results can be presented by using a matrix as in the figure below (Fig. 4.3):

In Graph 1, materiality of topics has been prioritized on the bases of impact and financial materiality scores.

In Graph 2, materiality of topics has been assessed on the basis of scores over a predefined materiality threshold in both impact and financial materiality assessment (Fig. 4.4).

The objective of this final step is to synthesize the findings from the previous steps and generate a comprehensive list of material environmental, social, and governance (ESG) impacts, risks, and opportunities. This list will serve as the foundation for the development of the sustainability statement, which will communicate the organization's ESG performance and commitments to stakeholders.

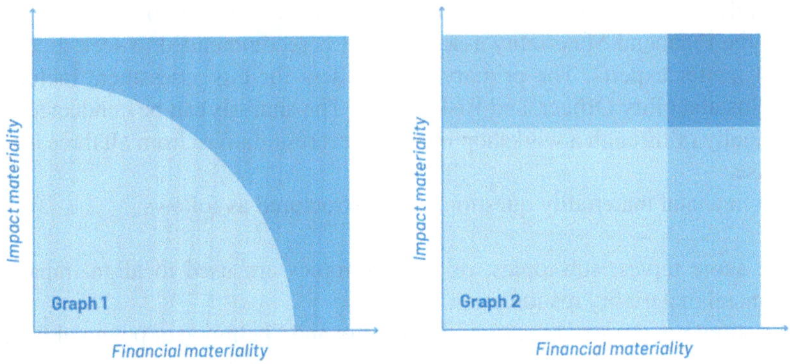

Fig. 4.3 Examples of materiality matrix—Source: Euronext Corporate Solutions, 2024 (Euronext Corporate Solutions (2024). The Ultimate guide to double Materiality Assessment)

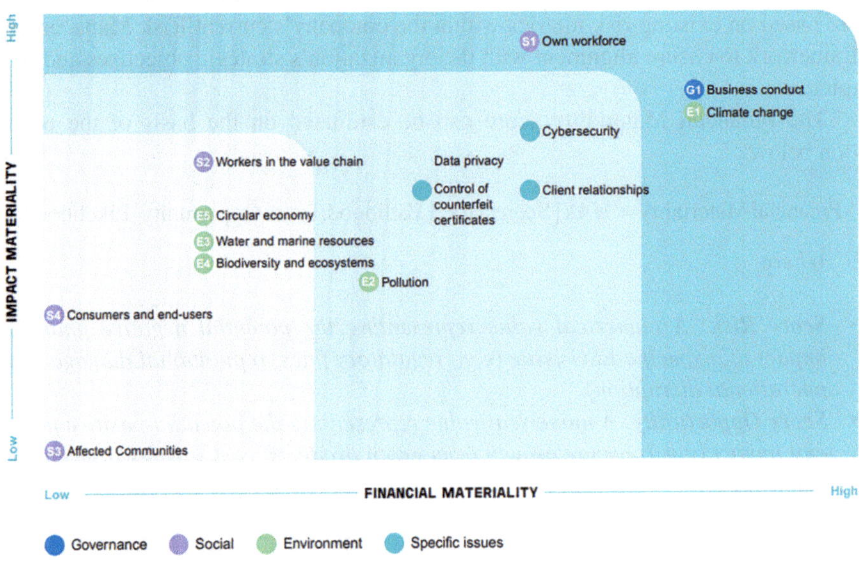

Fig. 4.4 Double Materiality Assessment (DMA) Example of Bureau Veritas—materiality threshold approach. Source: Bureau Veritas Integrated Report 2024, pg. 37

4.2 Stakeholder Mapping and Engagement

Research by García-Sánchez et al. (2021) reveals that stakeholder engagement through sustainable practices enhances brand equity and builds long-term trust.

Develop a set of clear and practical guidelines that the company can use to effectively communicate its ESG objectives to key stakeholders becomes, thus, decisive for a company.

These guidelines should aim to:

- **Promote transparency and accuracy** in the information provided. By adopting a clear and transparent approach to communicating sustainability objectives, the company can enhance stakeholder understanding and trust, demonstrating its genuine commitment to sustainability. Moreover, by aligning ESG objectives with corporate strategy, the company can convincingly demonstrate that sustainability is an integral part of its business model and long-term growth path.
- **Encourage active stakeholder engagement** in the process of defining and achieving ESG goals. Through an active stakeholder engagement approach, the company can build trusting relationships, gain a deeper understanding of their needs and expectations, and develop more effective and widely supported sustainability objectives.
- Emphasize the importance of **correctly identifying key stakeholders** and determining the most effective and efficient ways to communicate with them. The step of Stakeholder Mapping is a structured process that involves identifying, analyzing, categorizing, and positioning stakeholders based on the level of interest and influence they may have on the company's sustainability project.
- Highlight the need to assign clear responsibilities and foster continuous improvement in the **communication** of ESG objectives.

Stakeholder mapping and positioning is an important preliminary step since it affects the following strategy to be adopted to grant the effectiveness of the engagement process. A thorough understanding of stakeholder relationships, interests, and influence enables tailored engagement strategies that enhance project outcomes and reduce the risk of conflict or misalignment.

Stakeholder mapping and positioning—particularly through tools like the **Power-Influence Matrix**—constitutes a foundational step in designing effective engagement strategies. By systematically assessing stakeholders, based on their level of power and influence, organizations can tailor their communication and involvement approaches. This strategic alignment is critical to ensuring the success of engagement processes, as it allows for the prioritization of key actors.

The Power-Influence Matrix, a widely used stakeholder analysis tool, categorizes stakeholders into four quadrants based on their ability to influence outcomes and their interest or involvement in the issue. This framework helps identify:

- **Key players** (high power, high interest) who must be fully engaged;
- **Context setters** (high power, low interest) who need to be kept satisfied;
- **Subjects** (low power, high interest) who should be kept informed;
- **Crowd** (low power, low interest) who requires minimal communication effort (Fig. 4.5).

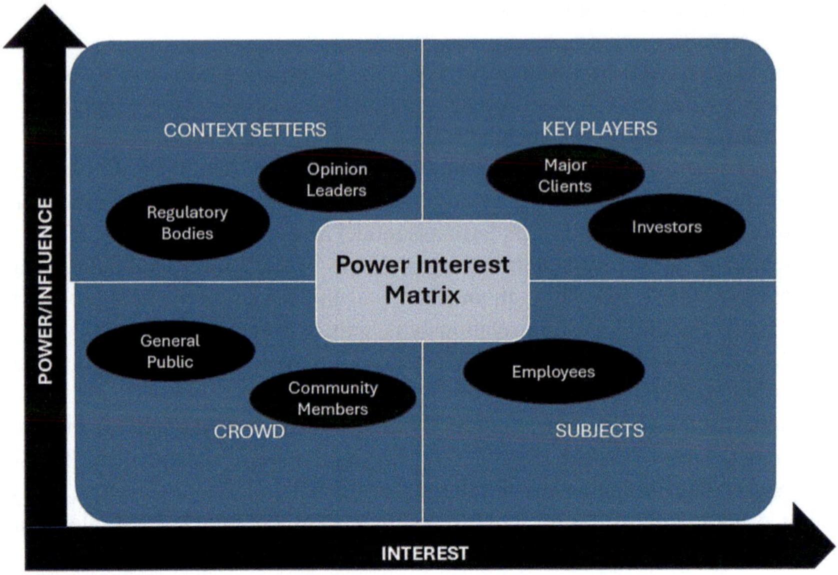

Fig. 4.5 Example of power-influence matrix. Source: Author Elaboration

EFRAG classified stakeholders into two groups:

(a) **Affected stakeholders**: These are individuals or groups whose interests are, or could be, positively or negatively impacted by the company's operations and its direct or indirect business relationships throughout the value chain.
(b) **Users of sustainability statements**: This group includes the primary users of general-purpose financial reports—such as current and potential investors, lenders, and other creditors (including asset managers, credit institutions, and insurance companies)—as well as other stakeholders who rely on sustainability disclosures. These may include business partners, trade unions, social partners, civil society organizations, NGOs, government bodies, analysts, and academics.

ESRS 2 requires organizations to be transparent about their engagement with affected stakeholders. This includes detailing any consultations with affected stakeholders to understand how they may be impacted.

According to Shift's guidance on stakeholder engagement in the ESRS (Shift Project, 2023), effective engagement must go beyond passive methods such as surveys or questionnaires. While these tools can provide useful quantitative insights, they often fail to capture the depth and nuance of stakeholder concerns—particularly those of individuals and communities directly affected by a company's operations. Shift emphasizes the importance of meaningful dialogue, which involves two-way communication, active listening, and the incorporation of stakeholder perspectives into decision-making processes.

Recommended methodologies for stakeholder engagement include:

- Focus groups and interviews with affected stakeholders;
- Workshops/Meetings involving internal and external experts;
- Grievance mechanisms and feedback channels;
- Collaborative forums with trade unions, NGOs, and community representatives.

FOCUS: Methodology for a Stakeholder Meeting Organization

The following methodology can be followed to organize a successful stakeholder meeting:

Planning: To ensure the meetings are held with the highest quality standards, an accurate planning will be designed so that all the organizational aspects are duly considered.

Good balance of representatives, coming from the different stakeholders groups should participate, to allow a well-balanced representativeness and to allow a good interaction during the meeting.

Creation of a dedicated webpage: A webpage specifically tailored to the meetings could help stakeholders to find relevant information on the event, its agenda and speakers as well as the scope of the materiality assessment. The use of free tools such as EventBrite for the creation of the meetings webpage could be useful.

Stakeholder engagement before the meetings, providing documentation, ensuring appropriate feedback is given to needs identified and different positions are represented during the meetings.

Appointment of moderators to ensure the participative engagement of stakeholders and stimulate the generation of ideas and the collection of inputs, avoiding one-man speeches and presentations.

Steps for Stakeholders Meetings Organization. Source: Author elaboration

During the meeting, the stakeholders are engaged in the double materiality assessment in the following ways:

- Assess the impact of the company's actions and activities on the identified potentially material topics.
- Evaluate the corresponding impact of the identified topics on the company's financial indicators (e.g., EBITDA, profit, net financial position).

Hence, stakeholders are asked to consider both ESG impacts and financial risks and opportunities, specifying whether the impact is positive, negative, or both, and assigning a score. These scores are used to define the threshold over which the topics are considered material (See Sect. 4.1 for details on the materiality assessment process).

4.3 Sustainability Reporting: The Structure and Some Examples

The origins of sustainability reporting can be traced back to environmental disclosures in the 1980s and 1990s. Over time, the scope expanded to include social and governance dimensions, reflecting the broader ESG framework.

A recent systematic review by Du Toit (2024)[6] highlights how sustainability reporting has matured over the past three decades, moving from voluntary, fragmented disclosures to more standardized and strategic communication tools. Robust sustainability reporting not only strengthens corporate reputation and fosters stakeholder trust, but also plays a critical role in driving long-term financial performance.

A sustainability report generally adheres to a well-defined structure aimed at promoting transparency, enhancing comparability, and ensuring alignment with internationally recognized reporting standards such as the Global Reporting Initiative (GRI), the European Sustainability Reporting Standards (ESRS), and the IFRS Sustainability Disclosure Standards (IFRS S1 and S2) (see Chap. 2). These frameworks provide a common language for organizations to communicate their environmental, social, and governance (ESG) performance in a way that is meaningful to stakeholders and aligned with regulatory expectations.

The present chapter seeks to outline the typical structure of a sustainability report, offering practical guidance for companies and consultants engaged in the reporting process.

Usually, a sustainability report starts with an **Executive Summary / CEO Statement** to describe the company's high-level overview of sustainability achievements, challenges, and strategic direction.

Follows a **general presentation of the company** in terms of number of employees, geographical locations, business model and operations, the supply chain, key markets, and the sustainability vision, i.e. a sort of strategic and aspirational statement that outlines an organization's long-term commitment to sustainable development and serves as a guiding principle for integrating environmental, social, and governance (ESG) considerations into the company's core business strategy. In this introductory part, it is also important to define which framework (e.g., GRI, CSRD, IFRS) the Report is aligned to.

A primary chapter in any sustainability report is the one dedicated to the **Materiality Assessment**, as it serves as the cornerstone for identifying the key Environmental, Social, and Governance (ESG) issues that are most relevant to both the company and its stakeholders. Through this process, the company systematically determines which topics are material—meaning they have significant economic, environmental, or social impacts or substantively influence stakeholder decisions (See Sect. 4.1).

[6] Du Toit E. (2024). Thirty Years of Sustainability Reporting: Insights, Gaps and an Agenda for Future Research Through a Systematic Literature Review. Sustainability, 16(23)

The Materiality assessment typically includes a detailed account of the stakeholder engagement process, which is essential for ensuring that the materiality assessment reflects a broad spectrum of perspectives. As highlighted by Kaspersen and Johansen (2023)[7], stakeholder engagement is not merely a procedural step but a critical mechanism that shapes the content and orientation of sustainability reports. It ensures that the report is not developed in isolation but is grounded in the expectations and concerns of those affected by the company's operations (See the previous Sections for details).

The chapter must also present the list of material topics, often visualized in a materiality matrix, and explains their relevance and prioritization. This transparency enhances the credibility of the report and supports strategic alignment with sustainability goals. Finally, the materiality assessment is increasingly seen as a dynamic and iterative process, evolving with stakeholder expectations, regulatory changes, and emerging risks. It is not only a reporting requirement but also a strategic tool for sustainability management.

Another important Chapter in any sustainability report is the one dedicated to **Governance and Ethics**. This section outlines the sustainability governance structure, clearly defining the roles and responsibilities of the board, executive leadership, and sustainability committees. It also addresses the company's commitment to ethical conduct, regulatory compliance, and an effective risk management.

Effective governance is increasingly recognized as a critical enabler of credible ESG performance. As Robertson et al. (2021)[8] argue, strong governance frameworks—characterized by board diversity, robust compliance systems, and transparent reporting lines—are essential for managing ESG risks and ensuring long-term value creation. The Governance and Ethics Chapter often includes also mechanisms for ethical oversight, such as codes of conduct, whistleblower protections, and anti-corruption policies. These elements are vital for maintaining stakeholders trust and mitigating reputational risks. Poiriazi et al. (2025)[9] emphasize the positive correlation between weak governance structures and greenwashing thus undermining the credibility of sustainability disclosures.

This alignment between governance and sustainability strategy is not only a best practice but a growing expectation from investors, regulators, and civil society.

The environmental and social performance topics deserve, as well, a dedicated Chapter.

[7] Kaspersen, M. and T.R. Johansen (2023) *Stakeholder Engagement and Materiality Assessments in Sustainability Reporting* in Measuring Sustainability and CSR: From Reporting to Decision-Making. Springer, Part of the book series: Ethical Economy (SEEP,volume 64)

[8] Robertson E., S. Hopkins and S. Toms (2021) Sustainability and ESG: The Governance Factor and What It Means for Businesses. Harvard Law School Forum on Corporate Governance

[9] Poiriazi, E., Zournatzidou, G., Konteos, G., & Sariannidis, N. (2025). Analyzing the interconnection between environmental, social, and governance (ESG) criteria and corporate corruption: Revealing the significant impact of greenwashing. Administrative Sciences, 15(3), 100. https://doi.org/10.3390/admsci15030100

In **the Environmental Performance Chapter**, a central component should be the company's climate change mitigation and adaptation strategy, which outlines both the long-term vision and the short-term actions taken to reduce greenhouse gas (GHG) emissions and build resilience against climate-related risks. Equally important is the quantification of the company's carbon footprint, encompassing Scope 1, 2, and, where feasible, Scope 3 emissions, in accordance with the GHG Protocol (See Sect. 6.2). This data should be contextualized with intensity metrics and year-over-year comparisons to illustrate progress and identify areas for improvement. The company, however, should also present data about the energy consumption, waste management and recycling, and water management.

An increasingly critical area of focus is the impact on biodiversity, particularly for companies operating in, or sourcing from, ecologically sensitive areas. Reporting should include assessments of ecosystem impacts and mitigation measures, potentially guided by tools such as the Global Biodiversity Score[10] or the Science Based Targets for Nature.[11]

Furthermore, the Environmental Performance Chapter should highlight the company's efforts toward a circular economy, which is gaining recognition as a systemic solution to environmental degradation and climate change. Circular strategies—such as product life extension, resource recovery, and closed-loop systems—can significantly reduce environmental footprints and foster innovation.

Social performance is another critical pillar of sustainability. The dedicated Chapter should reflect how the organization manage relationships with employees, suppliers, communities, and broader society. It encompasses both internal practices—such as labor conditions and diversity—and external impacts, such as community development and human rights. Among the key themes typically addressed in the chapter of a sustainability report dedicated to social matters are employee well-being, diversity, equity, and inclusion (DEI), respect for human rights and responsible labor practices, as well as community engagement and broader social impact. These topics reflect the organization's commitment to fostering a fair, inclusive, and socially responsible environment, both within its operations and in its interactions with external stakeholders.

Employees' well-being includes occupational health and safety, work–life balance, mental health support, and fair compensation. Reporting on these aspects demonstrates a company's commitment to human capital and long-term productivity.

Diversity, equity, and inclusion (DEI) reporting has become a critical area of focus, particularly in light of global social justice movements and investor

[10] The Global Biodiversity Score (GBS) is a corporate biodiversity footprint assessment tool: it can be used to evaluate the impact as well as the dependencies of companies and investments on biodiversity. Available at: https://tnfd.global/tools-platforms/global-biodiversity-score/

[11] This first release from SBTN equips companies to assess their environmental impacts and set targets beginning with freshwater and land enabling companies to both reduce their negative impacts and increase positive ones for nature and people. Available at: https://sciencebasedtargetsnetwork.org/how-%20it-works/the-first-science-based-targets-for-nature/

expectations. Berardi et al. (2025)[12] emphasize that DEI disclosures are increasingly linked to governance quality, stakeholder trust, and organizational legitimacy. Key indicators include workforce demographics, inclusive hiring practices, and board diversity measured through specific metrics (gender, age, ethnicity, disability, compensation).

Regarding Human Rights and Labor Practices, companies are expected to report on their adherence to international labor standards, prevention of forced and child labor, and respect for freedom of association. These disclosures are often aligned with frameworks such as the UN Guiding Principles on Business and Human Rights[13] and the ILO conventions.

Finally, the area of community engagement and social impact covers corporate philanthropy, local economic development, education initiatives, and partnerships with civil society. Effective reporting includes both qualitative narratives and quantitative impact metrics, demonstrating how the organization contributes to the well-being of the communities in which it operates.

Economic performance is a critical dimension to address within a sustainability report, as it underpins a company's capacity for long-term value creation and financial resilience. A robust economic foundation enables organizations not only to sustain their operations but also to invest in initiatives that generate *shared value*—benefits that extend beyond the firm to society at large.

The concept of shared value, introduced by Michael E. Porter and Mark R. Kramer (2011)[14], refers to policies and operating practices that enhance the competitiveness of a company while simultaneously advancing the economic and social conditions in the communities in which it operates. This approach redefines the role of business in society, emphasizing that sustainable innovation and responsible sourcing can simultaneously drive profitability and societal progress.

A Sustainability Report should, for all of these areas, define targets in the short, medium, and long term, and a structured system of Key Performance Indicators (KPIs) to monitor progress in time and against goals. These KPIs can be quantitative or qualitative and define the roadmap the company has decided to follow to achieve future targets.

KPIs should be directly linked to material issues to drive meaningful change. Moreover, integrating KPIs with recognized frameworks such as the UN Sustainable

[12] Berardi, L., Mook, L., Monaco, M.L. (2025). The Reporting of Diversity, Equity, and Inclusion by Nonprofit Organizations. First Evidence from the Literature. In: Agostini, M., Beretta, V., Demartini, M.C., Ghio, A., Trucco, S. (eds) Diversity and Equity in Accounting. SIDREA Series in Accounting and Business Administration. Springer, Cham. https://doi.org/10.1007/978-3-031-78247-3_3

[13] United Nations. (2011). Guiding Principles on Business and Human Rights: Implementing the United Nations "Protect, Respect and Remedy" Framework. Office of the High Commissioner for Human Rights. https://www.ohchr.org/en/publications/reference-publications/guiding-principles-business-and-human-rights

[14] Porter, M. E., and M. R. Kramer. "Creating Shared Value." Harvard Business Review 89, nos. 1–2 (January–February 2011): 62–77.

Development Goals (SDGs) enhances credibility and alignment with global sustainability standards.

All main sustainability reporting frameworks such as the Global Reporting Initiative (GRI), the International Financial Reporting Standards for Sustainability (IFRS), and the European Sustainability Reporting Standards (ESRS) emphasize the importance of setting measurable objectives and tracking performance through relevant indicators. For example:

- **Environmental KPIs** may include greenhouse gas emissions (Scope 1, 2, and 3), energy consumption, water usage, and waste reduction.
- **Social KPIs** often track employee turnover, diversity and inclusion metrics, health and safety incidents, and community engagement.
- **Governance KPIs** can include board diversity, ethics and compliance training, and risk management effectiveness.

The integration of KPIs into sustainability strategy is not only a best practice but a necessity for demonstrating accountability and driving continuous improvement.

It is important to underline that the report has to be verified by an external party which releases an independent assurance report that requires a dedicated chapter in the sustainability report. The external party evaluates and verifies the credibility of non-financial information, such as sustainability data, to ensure it meets certain standards (e.g., ISAE 3000, AA1000AS).

Focus

An Example of Sustainability Report According to ESRS: UNILEVER

The Unilever Annual Report and Accounts 2024[15] is composed of four integrated reports: The Strategy Report, The Governance Report, the Financial Statements, and the Sustainability Statements. This structure reflects a growing trend among leading corporations to integrate sustainability into core business reporting, aligning with the principles of integrated reporting as advocated by the International Integrated Reporting Council (IIRC).

This Box will describe only the Sustainability Statement Part that is structured as follows:

General Information;
Environmental Disclosures;
Social Disclosures;
Governance Disclosures;
Independent Assurance Report.

[15] https://www.unilever.com/files/unilever-annual-report-and-accounts-2024.pdf

This format aligns with global best practices in sustainability reporting, such as those outlined by the Global Reporting Initiative (GRI) and the European Sustainability Reporting Standards (ESRS), which emphasize the importance of transparency across ESG dimensions.

The General Information Section is structured in 6 Chapters: (i) General Basis for Preparation; (ii) Governance; (iii) Strategy and Business Model; (iv) Interest and View of Stakeholders; (v) Double Materiality; (vi) Policies Adopted to Manage Sustainability Matters.

The section starts with a "**General Basis for Preparation**" Chapter, where it is first reported the declaration that Unilever PLC has prepared the report in accordance with the European Sustainability Reporting Standards (the ESRS) as issued by Delegated Regulation (EU) 2023/2772 on 31 July 2023 (first year of reporting according to ESRS) and it is described the scope of the report. The section summarizes the basis of preparation for the sustainability statement, including the governance of the company's sustainability strategy and the assessment of material impacts, risks, and opportunities (IROs).

The boundaries of the disclosures are described declaring that they include the upstream and downstream value chain, to the extent that they are connected to Unilever's material impacts, risks, and opportunities. For the sake of clarity, the upstream value chain includes all suppliers involved in providing raw materials and packaging for Unilever's products. The company works with over 50,000 direct suppliers globally and also considers subcontractors and third-party manufacturers as part of this chain. Downstream value chain, instead, covers the network of distributors, retailers, and other partners who help deliver Unilever's products to consumers. It includes both large and small businesses, as well as agents, franchisers, and importers who sell or distribute on Unilever's behalf.

In the Chapter, it is also highlighted how and when comparisons with past data are made and how goals are set and tracked. Regarding the latter, it is important to explain how targets are developed: (i) Bottom-up planning; (ii) Industry benchmarks; (iii) Strategic ambition.

Bottom-up planning: Unilever integrates sustainability targets directly into the financial growth plans of its five Business Groups. This approach ensures that goals are grounded in operational realities and informed by granular data, such as product-level carbon footprints and supplier-specific emissions.

Industry benchmarks: The company aligns its targets with recognized standards, including the Science Based Targets initiative (See Chap. 6.2.1 for details about SBTi), to ensure credibility and comparability. For example, Unilever's near-term Scope 3 GHG reduction targets are validated by the SBTi and benchmarked against sectoral pathways consistent with the Paris Agreement.

Strategic ambition: Unilever's overarching goal of achieving net zero emissions across its value chain by 2039 reflects a long-term strategic vision. This ambition drives innovation, stakeholder engagement, and systemic change, as seen in initiatives like the Supplier Climate Programme and advocacy for policy reform.

The "General Basis for Preparation" Chapter describes also how data that cannot be directly measured have been handled, defines how metrics are calculated (using ESRS or company-specific definitions), explains estimation methods, discloses uncertainty (key assumptions, judgements, and approximations).

In the General Information Section, Unilever has also included a **"Governance" Chapter** where it is defined who is responsible for sustainability at the highest level. Unilever clarifies that the Board holds ultimate accountability, identifies the CEO and executive leadership team as responsible for operational oversight and links sustainability governance to the broader corporate governance framework.

In particular, Unilever lists relevant Board subcommittees and their roles in contributing to sustainability governance:

1. Corporate Responsibility Committee: In charge of managing strategy and performance. The Committee also reviews sustainability-related impacts, risks, and opportunities.
2. Audit Committee: Responsible for risk management and disclosure assurance.
3. Compensation Committee: Supports the delivery of the sustainability strategy through the alignment of Unilever's long-term incentive plan.
4. Nominating and Governance Committee: Responsible for the definition of the board composition and expertise.

The Governance Chapter also describes the management bodies and how sustainability is embedded in day-to-day operations, how sustainability performance is linked to rewards, how the company identifies and manages sustainability risks describing also due diligence processes for human rights and environmental impacts and the sustainability reporting controls (processes to assess and manage risks related to the integrity of the information disclosed in the sustainability statement).

The "**Strategy and Business Model**" Chapter describes the company strategy and business model i.e. the company value proposition, target markets, cost structure, and revenues streams.

The **"Interest and View of Stakeholders"** chapter describes the engagement processes and identifies six stakeholder groups as critical to the future success of the company: shareholders, employees, consumers, customers, suppliers & business partners, and planet & society.

A pivotal Chapter is referred to the **Double materiality approach description**. The Chapter, after an introduction on the principle of double materiality, described the step-by-step methodology used to identify material topics:

1. Identification of potential IROs gathering inputs from stakeholder engagement, internal experts, existing risk assessments.
2. Impact Materiality Assessment estimating a score based on severity and likelihood.
3. Financial Materiality Assessment of risks and opportunities, Scenario analysis and magnitude and likelihood. Identifying risks and opportunities needs to link ESG issues to business performance and describe the interaction of material impacts and risks with strategy and business model also by developing a scenario analysis.
4. Validation and disclosure by mapping validating findings with senior leaders and sustainability experts and mapping material IROs to ESRS disclosure requirements.

At the end of the Chapter, a summary of material IROs is included in a table and divided by ESG area.

Finally, the "**Policies Adopted to Manage Sustainability Matters**" chapter shows how principles, policies, and governance mechanisms guide responsible behavior and ensure compliance across the organization and its partners. It introduces the core ethical and behavioral standards and shows how policy implementation is managed and monitored.

The **"Environmental Disclosures" Section** has the aim of communicating how the company manages and reports its environmental impacts, risks, and opportunities. It defines the scope of environmental disclosures and links to Unilever's four sustainability priority areas: Climate, Nature, Plastics, and Livelihoods. It also clarifies the alignment with ESRS requirements and Unilever's broader sustainability strategy. It is important, in fact, to begin with a strategic overview that connects environmental goals to business priorities, this approach helps stakeholders see sustainability not as a separate initiative, but as an integral part of long-term value creation and risk management.

The Chapter continues organizing disclosures by thematic areas. Each environmental topic is addressed in its own subsection, typically following this format:

a. Climate Change (ESRS E1)
b. Pollution (ESRS E2)
c. Water and Marine Resources (ESRS E3)
d. Biodiversity and Ecosystems (ESRS E4)
e. Resource Use and Circular Economy (ESRS E5)

For each subsection, the report define targets and the relative timeline, the interaction of material impacts and risks with strategy and business model and the policies adopted to manage related IROs. Key actions taken to mitigate risks or to reach targets are also described. Metrics used to monitor progress in reaching the targets have to be clearly explained and data should be reported in dedicated tables showing the performance with respect to, at least, the previous year.

The subsection ends with the EU Taxonomy Disclosures KPIs reporting Turnover, Opex, and Capex KPIs.

The "Social Disclosures" Section has a structure similar to the previous one (disclosures organized by thematic area) but a subchapter dedicated to the company's approach to human rights and the commitment to respect human rights extends across operations and value chain. The Human Rights Due Diligence process is also described. Even in this Chapter, each social topic is addressed in its own subsection and for each subsection targets and the relative timeline are reported as well as the interaction of material impacts and risks with strategy and business model and the policies adopted to manage related IROs. Key actions taken to mitigate risks or to reach targets are also described, as well as the metrics used to monitor progress in reaching the targets.

The same structure is followed also for **the Governance Disclosures Section.** The role of administrative, management, and supervisory bodies is described as well as the list of material IROs reported for the ESRS G1 regarding to business conduct. Business conduct policies and corporate culture are presented, as well as the metrics and targets used.

The Report concludes with the **Independent Assurance Report.**

References

Berardi, L., Mook, L., & Monaco, M. L. (2025). The reporting of diversity, equity, and inclusion by nonprofit organizations. First evidence from the literature. In M. Agostini, V. Beretta, M. C. Demartini, A. Ghio, & S. Trucco (Eds.), *Diversity and equity in accounting. SIDREA series in accounting and business administration*. Springer. https://doi.org/10.1007/978-3-031-78247-3_3

Calabrese, A., Costa, R., Levialdi Ghiron, N., & Menichini, T. (2019). Materiality analysis in sustainability reporting: A tool for directing corporate sustainability towards emerging economic, environmental and social opportunities. *Technological and Economic Development of Economy, 25*(5), 1016–1038. https://doi.org/10.3846/tede.2019.10550

Dewi, A. A., Saraswati, E., Rahman, A. F., & Atmini, S. (2023). Materiality, stakeholder engagement disclosure, and corporate governance: Critical elements for the quality of sustainability reporting. *Cogent Business & Management, 10*(1), 2175437. https://doi.org/10.1080/23311975.2023.2175437

Du Toit, E. (2024). Thirty years of sustainability reporting: Insights, gaps and an agenda for future research through a systematic literature review. *Sustainability, 16*(23).

EFRAG. (2022). *ESRS1 general requirements*.

Euronext Corporate Solutions. (2024). *The ultimate guide to double materiality assessment*.

García-Sánchez, I. M., Hussain, N., Khan, S. A., & Martinez-Ferrero, J. (2021). Sustainability and corporate social responsibility disclosure: Effects on brand equity. *Business Strategy and the Environment, 30*(3), 1410–1423. https://doi.org/10.1002/bse.1234

Kaspersen, M., & Johansen, T. R. (2023). *Stakeholder engagement and materiality assessments in sustainability reporting in measuring sustainability and CSR: From reporting to decision-making* (Ethical economy (SEEP, volume 64)). Springer.

Poiriazi, E., Zournatzidou, G., Konteos, G., & Sariannidis, N. (2025). Analyzing the interconnection between environmental, social, and governance (ESG) criteria and corporate corruption: Revealing the significant impact of greenwashing. *Administrative Sciences, 15*(3), 100. https://doi.org/10.3390/admsci15030100

Robertson, E., Hopkins, S., & Toms, S. (2021). *Sustainability and ESG: The governance factor and what it means for businesses*. Harvard Law School Forum on Corporate Governance.

Shift Project. (2023). *Stakeholder engagement in the ESRS*. https://shiftproject.org/wp-content/uploads/2023/08/Stakeholder-Engagement-in-the-ESRS.pdf

United Nations. (2011). *Guiding principles on business and human rights: Implementing the United Nations "protect, respect and remedy" framework*. Office of the High Commissioner for Human Rights. https://www.ohchr.org/en/publications/reference-publications/guiding-principles-business-and-human-rights

How to Define a Strategic Action Plan for Sustainability

5

A Strategic Action Plan for Sustainability serves as a structured roadmap that guides organizations in the systematic development, implementation, and communication of their sustainability commitments. It is a management and planning tool and it is forward-looking and operational in nature, designed to guide the implementation of sustainability goals across the organization. These plans typically include defined targets, timelines, responsible units, performance indicators, and mechanisms for monitoring and review.

The definition of a Strategic Action Plan for Sustainability requires 6 steps (Fig. 5.1):

The role of a strategic consultant in this context is to guide the organization through this comprehensive process. This begins with an assessment of the company's ESG maturity and awareness, its current sustainability planning, reporting, and communication practices. It involves analyzing the structure and effectiveness of the company's ESG governance system, including how sustainability is integrated into decision-making processes and how responsibilities are distributed across the organization.

A critical component of this analysis is the evaluation of the company's value chain. This includes assessing the level of ESG awareness and reporting among suppliers and partners, as well as the degree to which the company engages its value chain in sustainability initiatives. Understanding the maturity and responsiveness of the value chain is essential for identifying potential risks and opportunities, as well as for ensuring that sustainability efforts are consistent and credible across the entire business ecosystem.

Equally important is the identification and engagement of stakeholders. The consultant works closely with the company to map its stakeholders, understand their expectations, and evaluate their level of engagement with the company's ESG initiatives.

M. Gussoni, *Integrating Strategic Sustainability Practices in Business*, CSR, Sustainability, Ethics & Governance,
https://doi.org/10.1007/978-3-032-09617-3_5

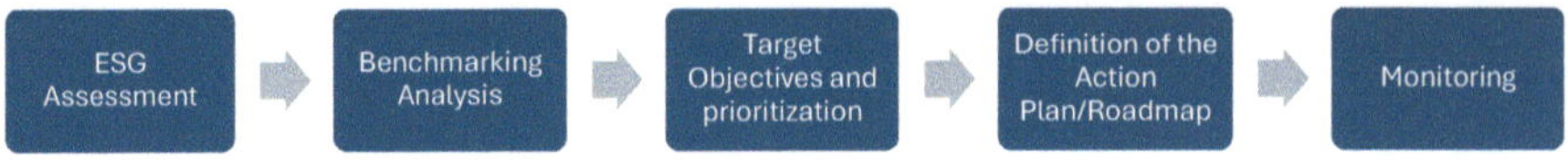

Fig. 5.1 Six steps to define a strategic action plan for sustainability. Source: Author Elaboration

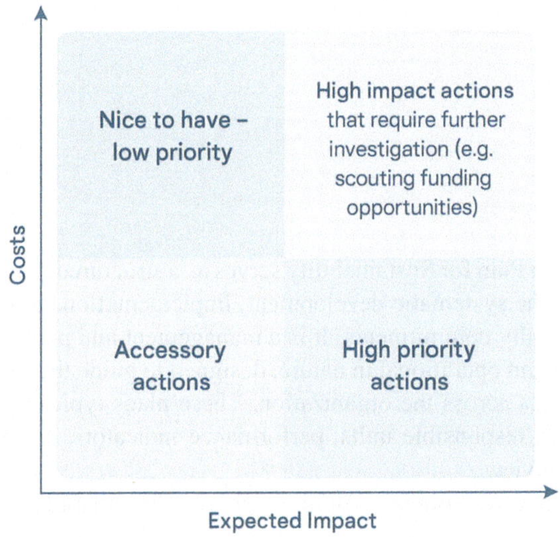

Fig. 5.2 Actions prioritization matrix example. Source: Author Elaboration

The consulting process also includes a detailed analysis of the company's ESG impacts, risks, and opportunities and a benchmarking with peers. This involves both qualitative and quantitative assessments, drawing on internal data, industry benchmarks, and regulatory frameworks such as the European Sustainability Reporting Standards (ESRS). The goal is to provide a clear and evidence-based understanding of the sustainability challenges and opportunities facing the organization.

After the benchmarking analysis, the company will be able to understand its positioning with respect to competitors and prioritize target and objectives in order to reduce the identified weaknesses relative to specific KPIs, e.g. higher energy consumption than the peers average. A wide set of actions could be usually deployed to improve performance or reach defined targets.

The prioritization of these actions is a critical step and become the base for the definition of the strategic action plan, which outlines the priority areas for intervention and sets the course for achieving the company's desired sustainability positioning in alignment with regulatory expectations and stakeholder demands. Often the prioritization of possible actions is based on a cost/impact assessment and could be the results of a company's top-management consultation (See Fig. 5.2 above).

After the prioritization process, the improvement actions identified as necessary to achieve the objectives and targets (High Priority Actions) may also be organized

Table 5.1 RACI matrix example

ESG Area	Action ID	Business Unit A	Business Unit B	Function C
E	Action 1	R	A	C
	Action 2	R	A	C
	…	…	…	…
S	Action 3	R	C	I
	Action 4	R	C	I
	…	…	…	…
G	Action 5	I	R	I
	Action 6	I	R	I
	…	…	…	

Notes: *R* Responsible: The person or people who actually do the work to complete the task. They are responsible for action and implementation; *A* Accountable: The person who is ultimately answerable for the correct and thorough completion of the task. This includes delegating the work and approving it. There should only be one Accountable person per task; *C* Consulted: People who provide input, advice, or expertise necessary for completing the task. This is typically a two-way communication; *I* Informed (just to complete the acronym): People who need to be kept up to date on progress or decisions, but do not need to be consulted or take action
Source: Author Elaborations

by ESG areas and by Business Unit/Functions in order to assign specific responsibilities in their implementation. Usually, a strategic tool used to define the responsibilities within an organization to take action and reach defined targets is the RACI matrix (See Table 5.1).

According to their level of priority, actions can be also planned in a roadmap over a multi-year time horizon (GANTT), with annual targets, aligning the Strategic Plan with the Industrial Plan, if available.

It is important that, for each objective, the following has been finally defined:

- Baseline year;
- Target;
- Timeline;
- Activities;
- Responsibilities;
- Resources and timing (financial timeline);
- Cost estimate.

The cost estimate is generally carried out based on literature, requests for quotations, and expert evaluations.

5.1 Data Collection Process for ESG Assessment

The ESG assessment process typically begins with the collection of data, information, and documentation related to the company's past and current sustainability efforts. This includes reviewing existing reports, accounting records, and communication materials. The aim is to build a comprehensive picture of the company's sustainability journey to date and to identify any gaps or areas for improvement.

Data collection may be conducted through focus groups involving top management, which allow for an in-depth evaluation of ESG management and reporting procedures. Alternatively, or in addition, a structured questionnaire may be used. This questionnaire should be developed based on the requirements of the ESRS, the provisions of the CSRD, previous consulting experiences, and recognized best practices in the field.

In collaboration with the sustainability and human resources departments—or directly with senior management—a stakeholder mapping exercise can be carried out to identify and categorize the company's key stakeholders. Similarly, in cooperation with the sustainability and procurement functions, the consultant can examine the company's current and planned initiatives for engaging the value chain in ESG actions. Once priority suppliers are identified, a targeted questionnaire has to be administered to assess their level of ESG awareness and reporting.

Upon completion of these activities, the consultant delivers a comprehensive evaluation that includes the organization's **ESG maturity level**, the effectiveness of its sustainability management and communication, the engagement level of its value chain, the mapping and involvement of stakeholders, and the relevance of sustainability topics to the company's operations. This also includes a detailed analysis of the ESG impacts, risks, and opportunities the company must address.

5.2 Monitoring

Once an implementation roadmap for the Strategic Action Plan has been defined, it is also important to establish a monitoring system to track the achievement of the set objectives and to identify any deviations from the planned timelines and costs.

To this end, a dashboard should be structured with all the defined KPIs, in order to have an overall view of their progress. A data collection model must also be prepared for calculating these KPIs.

The Data Collection Model should include, at least, the following elements:

- The specific data sources (e.g., energy consumption monitoring systems, electricity bills, paper consumption, etc.);
- In the absence of primary data, the specific calculation methods based on authoritative and scientifically grounded sources (e.g., calculation of average per capita water consumption; calculation of average production of waste similar to municipal waste, etc.);
- Data collection timelines;
- Methods for collecting, analyzing, cataloging, and archiving evidence;
- Guidelines for the annual planning of data collection activities.

The KPI dashboard should also be structured based on the reference standards for sustainability reporting. For each KPI, the following elements should be defined:

- Reference standard;
- Input data and elements for KPI calculation;
- Formula for KPI calculation;
- Monitoring methods (tools and timelines);
- Intermediate and final control methods.

Establishing a structured and consistent system for monitoring the implementation of the Strategic Action Plan is essential to ensure transparency, accountability, and continuous improvement. The integration of a comprehensive KPI dashboard—aligned with recognized sustainability reporting standards—enables timely tracking of progress, early identification of deviations, and informed decision-making.

By defining clear data sources, standardized calculation methods, and robust data collection and control procedures, organizations can ensure the reliability and comparability of performance indicators over time. This approach not only supports internal governance but also strengthens external communication and stakeholder engagement, reinforcing the organization's commitment to sustainability and strategic excellence.

Tools and Methods for Strategic Consultants in Corporate Decarbonization

6

This chapter will present tools and methods useful to the strategic consultant to support companies in the decarbonization process.

According to the EDGAR report 2024 (Crippa et al., 2024), global greenhouse gas (GHG) emissions, in 2023, reached an unprecedented 53.0 gigatonnes of CO_2 equivalent (Gt CO_2eq), excluding emissions from Land Use, Land-Use Change and Forestry (LULUCF). This marks the highest level ever recorded, with a year-on-year increase of 1.9%, equivalent to 994 million tonnes of CO_2eq compared to 2022.

The rise in emissions was primarily driven by increases in fossil CO_2, which accounted for 73.7% of total GHG emissions. Other contributors included methane (CH_4) at 18.9%, nitrous oxide (N_2O) at 4.7%, and fluorinated gases (F-gases) at 2.7%.

Among the top emitters—China, the United States, India, the EU27, Russia, and Brazil—China recorded the largest absolute increase in emissions (+784 Mt. CO_2eq), while India saw the highest relative growth (+6.1%). Together, these six economies accounted for 62.7% of global GHG emissions, despite representing only 49.8% of the global population.

Given these premises and in the face of escalating climate risks and tightening regulatory frameworks, decarbonization has emerged as an environmental imperative. For strategic consultants, supporting companies in their journey toward net-zero emissions requires a robust toolkit of methods, frameworks, and sector-specific insights.

This chapter introduces the essential tools and approaches that consultants can deploy to guide organizations through the complex process of decarbonization, aligning sustainability goals with business value creation.

Decarbonization strategies typically begin with a comprehensive assessment of a company's carbon footprint. For this reason, Sects. 6.1 and 6.2, respectively, present the Life Cycle Assessment (LCA) Methodology and GHG calculation methods according to GHG Protocol. Once baseline emissions are established, consultants

M. Gussoni, *Integrating Strategic Sustainability Practices in Business*, CSR, Sustainability, Ethics & Governance,
https://doi.org/10.1007/978-3-032-09617-3_6

guide companies in setting science-based targets and developing tailored roadmaps. How to set Science-Based Targets is presented in Sect. 6.2.1.

Strategic consultants must also help clients navigating the GHG scopes by identifying high-impact areas (hot spots) and prioritizing interventions. Tools such as marginal abatement cost curves (MACC) are instrumental in this phase, as presented in Sect. 6.2.2.

Finally, Sect. 6.2.3 presents the carbon credit market which allows allowing entities, governments, or companies, to buy and sell carbon credits, each representing the right to emit one metric ton of carbon dioxide (CO_2) or its equivalent in other GHGs, to reach Net-Zero Targets by compensating residual CO2 emissions that cannot be reduced in other ways.

6.1 An Introduction to Life Cycle Assessment (LCA)

Life Cycle Assessment (LCA) is a methodology to compute the relevant emissions and the resources consumed for a product or a service, considering their entire life cycle, from "cradle to grave," i.e. from raw material acquisition or generation from natural resources, to final disposal.

This means that the calculation has to consider the extraction of the raw material used, the whole supply chain, the environmental impact of transportation, the product use, and the waste management.

The LCA allows, thus to:

- Compare the environmental impact of alternative products with similar functions taking into account their complete life cycle (comparative life cycle assessment);
- Identify hot spots, i.e. points in the life cycle where the relevant emissions are higher, this could suggest pattern for optimization and GHG emission reduction;
- Assess the impact of a technological innovation in the entire life cycle, sometimes some innovations could reduce emissions at one point but could create negative impacts in a decommissioning phase or regarding waste management.

The framework for LCA is provided by the ISO 14040 and 14044 standards. Moreover, a technical guide to these standards is provided by the JRC (2010) "ILCD Handbook: General guide for Life Cycle Assessment—Detailed guidance. (source available at https://eplca.jrc.ec.europa.eu/uploads/ILCD-Handbook-General-guide-for-LCA-DETAILED-GUIDANCE-12March2010-ISBN-fin-v1.0-EN.pdf)" Figure 6.1 outlines the phases of the LCA according to ISO 14044-2018.

The "**Goal and Scope definition**" phase includes the definition of the boundaries of the study and the level of details which depends on the scope of the study and its following use. The criteria used to define boundaries have to be clearly explained in the LCA Report.

A useful way to represent boundaries of a LCA, for example, is a flow diagram showing all the units process and their interdependencies, from the acquisition of raw materials or intermediate products, through the transformation to the final product and its final use.

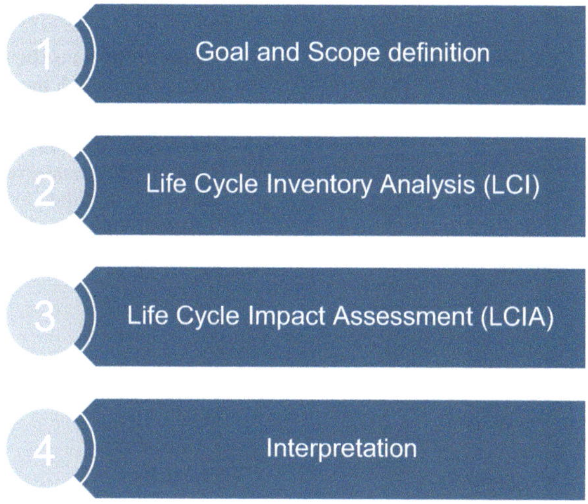

Fig. 6.1 The phases of the LCA according to ISO 14044-2018. Source: Author elaboration on the basis of ISO 14044-2018

The **Life Cycle Inventory (LCI) analysis** phase aims at the definition of an **input-output data model** within the defined boundaries. The collection of qualitative and quantitative data shall be performed for each unit process and all the data sources must be referenced. The inputs to be included in the model have to be chosen according to a cut-off criteria and on the basis of three different possible approaches:

1. **Mass**: In the model are included all the inputs that represent more than a certain percentage of the total mass input of the product system;
2. **Energy**: In the model are included all those inputs that represent more than a certain percentage of the total energy input of the product system;
3. **Environmental significance**: In the model are included all those inputs that represent more than a certain percentage of the total inputs of the product system selected for their environmental significance.

The three approaches can be used cumulatively to identify all the inputs that should be included in the study. Data to be included shall be based on a sensitivity analysis to determine their significance.

The collected data can be **secondary data** acquired from recognized environmental databases (e.g., ECOINVENT[1]) or, **primary data,** i.e. site-specific data collected, for each unit process, regarding material inputs, water consumption and energy inputs[2] and the relative material outputs (quantity) during a specified time period, by using a sampling approach (See: Table 6.1).

The golden rule is to describe the approaches in a detailed way to allow others to replicate the study.

[1] https://support.ecoinvent.org/introduction-to-the-database

[2] For example, heavy fuel oil, kerosene, natural gas, propane, coal, biomass, grid electricity, etc.

Table 6.1 Example of unit process datasheet

Unit process description:	**[A unit process is the smallest element in an LCA system for which data are collected.]**	
Sampling process time period:	**[A Sampling process approach requires to collect representative samples of data points (e.g., energy consumption, material use, emissions) and then use statistical methods to estimate the average and variability for the unit process.]**	
Material inputs:	**Quantity**	**Unit**
[...]	[...]	[...]
[...]	[...]	[...]
[...]	[...]	[...]
Water consumption:	**Quantity**	**Unit**
Drinking water	[...]	m^3
[...]	[...]	[...]
Energy inputs:	**Quantity**	**Unit**
Electricity	[...]	[...]
[...]	[...]	[...]
Material outputs:	**Quantity**	**Unit**
[...]	[...]	[...]
[...]	[...]	[...]

Source: Author elaboration on the basis of ISO 14044-2018

In the "**Life Cycle impact assessment (LCIA)" phase**, the environmental data collected during the LCI phase (e.g., emissions, resource use) are translated into environmental impact indicators. This helps assessing the potential environmental consequences of a product or system throughout its life cycle.

A good LCIA depends on the:

- Quality of collected data during the LCI phase. They should be sufficient to reach the scope of the study;
- A clear definition of the boundaries of the system and the data cut-off underlying rationale.

The three mandatory elements of the LCIA are:

1. Definition of the category endpoint(s) (areas of protection) and selection of impact categories (midpoint scores) and indicators.
2. The classification of the LCI results according to the selected impact indicators.
3. Calculation of the impact category indicators results (characterization).

The selection of the impact category indicators should be consistent with the goal and scope of the LCA study.

The first element regarding the definition of the Category endpoints describes the final environmental impacts the study is focused on (See Fig. 6.2 for an example) and the so-called Areas of Protection. These represent the final effects on areas of concern, e.g. human health, ecosystem quality (soil quality, biodiversity), natural resources, human-made environment.

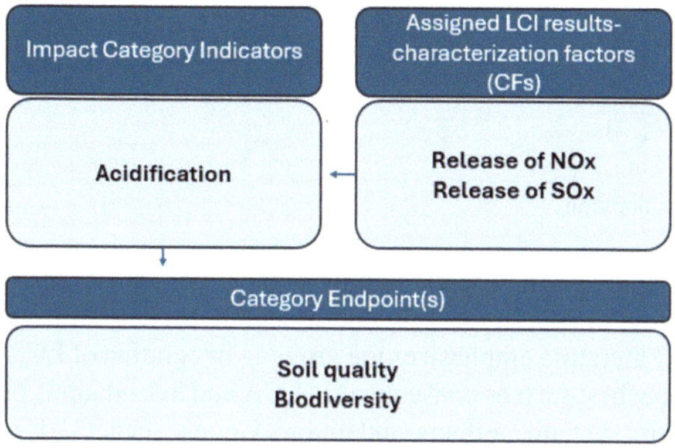

Fig. 6.2 Category indicators and endpoints example. Source: Author elaboration on the basis of ISO 14044

Most common impact category indicators are climate change, ozone depletion, eutrophication, acidification, human toxicity (cancer and non-cancer related), respiratory inorganics, ionizing radiation, ecotoxicity, photochemical ozone formation, land use, and resource depletion.[3]

LCI results could relate to one or more impact categories (e.g., SO_2 relates both to human health and acidification) and this has to be clearly explained. Variation in the quality of category indicators among the identified impact categories can influence the overall accuracy of the LCA; however, it is important to underline that a trade-off often exists between the simplicity and accuracy of the characterization model and this has to be taken into account and adequately balanced.

The last phase is the interpretation of the LCI and LCIA results according to the goal and scope of the LCA study. The interpretation shall include a sensitivity analysis on the significance of inputs, outputs, and methodological choices such as allocation rules and system boundary from the LCI and category indicators and models used in LCIA.

It is worth to underline that LCI results have to be interpreted considering they relate to inputs and outputs but not to environmental impacts. Only during the environmental impact assessment phase (Life Cycle Impact Assessment—LCIA), the multiple data relating to the input and output flows of the system, (inventory data) previously defined, are processed using appropriate environmental models in order to obtain a set of synthetic indicators (impact categories).

The results of the LCIA are usually computed and reported considering **1 unit of output** for each considered life cycle phase (e.g., raw material supply, transport, manufacturing, use, and end of life) See Table 6.2.

[3] European Commission-Joint Research Centre - Institute for Environment and Sustainability. (2011). International Reference Life Cycle Data System (ILCD) Handbook: Recommendations for Life Cycle Impact Assessment in the European context. First edition November 2011. EUR 24571 EN. Luxemburg: Publications Office of the European Union. ISBN 978-92-79-17451-3. DOI: 10.278/33030.

Table 6.2 Example of LCIA table for one output unit

Impact category	Unit	Raw material supply	Transport	Manufacturing	Total
Climate change	kg CO2 eq				
[...]	[...]				
Water use	m^3				

Source: Author elaboration

Life Cycle Assessment (LCA) is, thus, a cornerstone methodology for evaluating environmental impacts across product systems. It is worth to underline, however, that recent literature emphasizes the growing integration of LCA with dynamic modeling approaches, such as consequential LCA and hybrid input-output methods, to better capture systemic effects and indirect consequences (Finnveden et al., 2009). Furthermore, the emergence of temporal and spatial differentiation in inventory data (e.g., dynamic LCA) and the coupling of LCA with big data and AI-driven analytics represent promising avenues for improving accuracy and scalability (Beloin-Saint-Pierre et al., 2020). Open research lines include: (i) addressing data gaps and uncertainty propagation through probabilistic and machine-learning approaches; (ii) integrating social and economic dimensions into sustainability assessments via Life Cycle Sustainability Assessment (LCSA); and (iii) developing sector-specific guidelines for emerging technologies, such as hydrogen production and circular economy systems. These directions underscore the need for interdisciplinary collaboration and methodological innovation to ensure that LCA remains robust, transparent, and relevant in supporting sustainable decision-making in increasingly complex global value chains.

6.2 GHG Calculation Methods and Roadmaps to Decarbonization

Greenhouse Gas (GHG) Emissions are defined in Art. 3 of the Regulation (EU) 2018/842[4] as "*emissions in terms of tons of* CO_2 *equivalent of carbon dioxide* (CO_2)*, methane* (CH_4)*, nitrous oxide* (N_2O)*, hydrofluorocarbons (HFCs), perfluorocarbons (PFCs), nitrogen trifluoride* (NF_3) *and sulphur hexafluoride* (SF_6)*.*"

The GHG Protocol[5] equips consultants with a comprehensive set of tools and methodologies to help clients to develop accurate and transparent GHG emissions inventories defined as the list of emission sources and the associated emissions quantified using standardized methods.[6]

GHG Protocol tools support standardized reporting, enhance the credibility of climate disclosures, and provide a strategic foundation for tracking progress toward decarbonization goals and designing effective mitigation pathways.

[4] Regulation (EU) 2018/842 of the European Parliament and of the Council on binding annual greenhouse gas emission reductions by Member States from 2021 to 2030 contributing to climate action to meet commitments under the Paris Agreement and amending Regulation (EU) No 525/2013.

[5] https://ghgprotocol.org/

[6] https://www.epa.gov/

GHG Emissions are categorized into three types:

Scope 1 GHG Emissions are direct emissions that come from sources owned or controlled by the reporting entity. They result from activities such as fuel combustion in company-owned equipment or vehicles.

Scope 2 GHG Emissions are indirect emissions associated with the generation of purchased or acquired electricity, steam, heating, or cooling that the entity consumes. Although the emissions occur at the facility where the energy is produced, they are accounted for by the entity that uses the energy.[7]

Scope 3 GHG Emissions are indirect emissions not included in Scope 2, arising from activities across the entity's value chain, both upstream and downstream. They encompass a wide range of sources and are defined according to the Greenhouse Gas Protocol Corporate Value Chain (Scope 3) Standard (2011).

The referring standard for the GHG accounting and reporting is *The Greenhouse Gas Protocol: A Corporate Accounting and Reporting Standard (Revised Edition)* (Fig. 6.3).[8]

The GHG Protocol provides tools and guidelines for calculation of Scope 1, Scope 2, and Scope 3 emissions. These tools are typically electronic spreadsheets

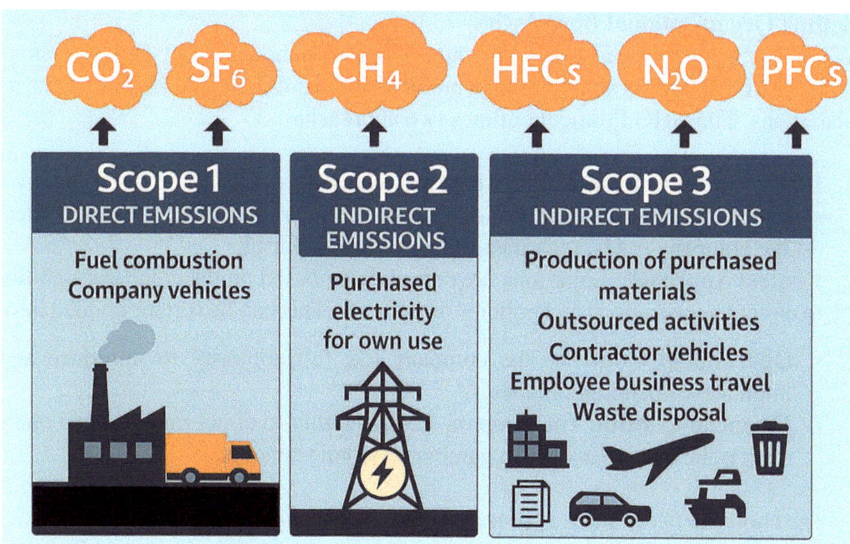

Fig. 6.3 Overview of scopes and emissions across a value chain. Source: Author elaboration of the basis of World Resources Institute & World Business Council for Sustainable Development. (2004). The Greenhouse Gas Protocol: A Corporate Accounting and Reporting Standard (Revised Edition), pg. 26

[7] International Sustainability Standards Board (ISSB). (2023). IFRS S2 Climate-related Disclosures. IFRS Foundation. Appendix A—Defined Terms, p. 21.

[8] World Resources Institute & World Business Council for Sustainable Development. (2004). The Greenhouse Gas Protocol: A Corporate Accounting and Reporting Standard (Revised Edition).

with accompanying step-by-step guidance documents. They can be grouped as follows:

- Cross-Sector Tools: Designed for broad applicability across various industries and business types, regardless of sector;
- Country-Specific Tools: Tailored to meet the unique needs and contexts of countries;
- Sector-Specific Tools: Primarily developed for particular industries or sectors, though they may also be adaptable to other contexts;
- Tools for Countries and Cities: Support national and local governments in monitoring and advancing progress toward their climate and sustainability targets.

The GHG inventory development process consists of four key steps:

1. Setting organizational and operational boundaries;
2. Identifying and calculating GHG emissions;
3. Ensuring inventory quality and consistency;
4. Set a GHG emission reduction target and track and report progress.

Setting Organizational Boundaries

When establishing organizational boundaries for greenhouse gas (GHG) emissions reporting, the company must choose a method for consolidating emissions across its operations. The GHG Protocol outlines two approaches:

1. **Equity Share Approach**: Emissions are accounted for based on the company's share of equity in each operation. This reflects the proportion of economic interest the company holds, regardless of operational control.
2. **Control Approach**: Emissions are consolidated based on the company's ability to direct the operations and policies of an entity. This can be further divided into:
   - **Operational Control**: The company has full authority to introduce and implement operating policies.
   - **Financial Control**: The company has the ability to direct financial and operating policies with a view to gaining economic benefits.

The choice between these approaches affects how emissions are attributed across joint ventures, subsidiaries, and partnerships. However, if a company **wholly owns and controls all of its operations**, the distinction between the equity share and control approaches becomes irrelevant—both methods will yield the same organizational boundary and emissions inventory.

This decision is foundational, as it determines which emissions sources are included in the company's GHG inventory and ensures consistency and transparency in reporting. Consultants supporting clients in this phase should thus gather details about: legal structure (subsidiaries, joint ventures, franchises), business operations (owned, leased, controlled assets), and geographic footprint. Moreover, it is important to clarify the client's reporting objectives (compliance, voluntary disclosure, internal management).

Setting Operational Boundaries

After an organization has defined its organizational boundaries in terms of operations it owns or controls, it has to set its operational boundaries.

An operational boundary delineates the range of both direct and indirect greenhouse gas emissions associated with the activities that fall within a company's defined organizational boundary. It helps determine which emission sources are included in the company's carbon footprint based on operational control, financial control, or equity share. If the company decides to include Scope 3 emissions, it has also to decide which categories to include (GHG Protocol lists 15 categories). However, the identification of relevant activities, and their categorization into scopes, is necessary also for Scope 1 and 2.

The established organizational and operational boundaries together constitute a company's inventory boundary.

Tracking Emissions Over Time

To enable meaningful and consistent tracking of emissions over time, companies must establish a reference point—known as the **base year emissions**—against which current emissions can be compared.

This baseline serves as a benchmark for measuring performance and progress. However, when significant structural changes occur, such as mergers, acquisitions, or divestments, it may be necessary to recalculate the base year emissions to maintain the integrity and comparability of emissions data over time.

While most companies designate a single year as their base year for emissions tracking, an alternative approach is to use an average of emissions over multiple consecutive years.[9] This method can help mitigating the impact of anomalies or irregular fluctuations in greenhouse gas emissions that might render a single year unrepresentative of typical operations.

Identifying and Calculating GHG Emissions

The first step to create an accurate accounting of companies' emissions is the identification of the emission sources. Appendix D of the Greenhouse Gas Protocol Accounting Standard[10] provides examples of Scope 1, Scope 2, and Scope 3 emissions sources by sector. It may be used as an initial guide to identify major GHG emission sources.

After sources identification, it is necessary to choose a calculation approach. The most common approach for calculating GHG emissions is through the application of **documented emission factors.** However, the 1996 IPCC Guidelines[11] outline a hierarchy of emission calculation approaches, ranging from the use of generic emission factors to direct monitoring of emissions.

[9] For instance, the U.K. Emissions Trading Scheme (ETS) uses the average emissions from 1998 to 2000 as its reference point for measuring reductions.

[10] World Resources Institute & World Business Council for Sustainable Development. (2004). The Greenhouse Gas Protocol: A Corporate Accounting and Reporting Standard (Revised Edition), pg. 92.

[11] IPCC (1996), Revised 1996 IPCC Guidelines for National Greenhouse Gas Inventories.

Table 6.3 Example of template for Scope 1 GHG emissions inventory

Emission source	Activity data	Units	Emission factor	EF units	Calculated emissions (tCO_2e)
Stationary combustion—natural gas	[...]	m^3	[...]	kg CO_2e/m^3	[...]
Mobile combustion—company vehicles	[...]	Liters	[...]	kg CO_2e/liter	[...]
Process emissions—manufacturing	[...]	kg of material	[...]	kg CO_2e/kg	[...]

Source: Author elaborations

For most small to medium-sized enterprises (SMEs) and many larger organizations, the calculation of greenhouse gas (GHG) emissions across Scopes 1, 2, and 3 typically follows standardized methodologies aligned with the Greenhouse Gas Protocol Corporate Standard.

Scope 1 emissions, which include direct GHG emissions from owned or controlled sources, are generally calculated using data on the quantities of commercial fuels used (e.g., natural gas, heating oil, diesel), process emissions (e.g., chemical reactions in manufacturing), and fugitive emissions. These emissions are quantified using published emission factors, often provided by national environmental agencies or international standards bodies. Where available, facility-specific or source-specific emission factors are preferred for greater accuracy (Table 6.3).

The generally used formula to compute Scope 1 GHG emissions is the following:

$$\text{GHG Emissions}\left(\text{tCO}_2\text{e}\right) = \text{Activity Data} \times \text{Emission Factor} \times \text{GWP}$$

Where:

- Activity Data: Quantity of fuel or material used (e.g., liters of diesel, kg of refrigerant)
- Emission Factor: Amount of GHG emitted per unit of activity (e.g., kg CO_2/liter of diesel)
- GWP (Global Warming Potential): Converts non-CO_2 gases (like CH_4 or N_2O) into CO_2 equivalents

Scope 2 emissions, which cover indirect emissions from the consumption of purchased electricity, steam, heating, or cooling, are primarily derived from energy consumption. These are then multiplied by emission factors that are either supplier-specific, based on local grid averages, or drawn from other reputable published sources.

Scope 2 emissions occur, thus, off-site, but are a consequence of the company's energy use. The GHG Protocol requires reporting Scope 2 using two methods:

Table 6.4 Example of template for Scope 2 GHG emissions inventory

Emission source	Activity data	Units	Emission factor	EF units	Calculated emissions (tCO_2e)
Purchased electricity	[...]	kWh	[...]	kg CO_2e/kWh	[...]
Purchased steam	[...]	MJ	[...]	kg CO_2e/MJ	[...]

Source: Author elaborations

1. Location-Based Method: Reflects the average emissions intensity of the grids where energy consumption occurs. The computing can be done by using national or regional grid average emission factors (e.g., from IEA, EPA, DEFRA).
2. Market-Based Method: Reflects emissions from electricity that companies have purposefully chosen (e.g., via renewable energy contracts or certificates) (Table 6.4).

Scope 3 emissions, which encompass all other indirect emissions occurring in the value chain (both upstream and downstream), are typically estimated using activity data such as business travel (e.g., passenger miles), fuel use by third parties, or purchased goods and services.

Scope 3 includes **15 categories** of emissions sources, such as:

Upstream:

1. Purchased goods and services.
2. Capital goods.
3. Fuel- and energy-related activities (not included in Scope 1 or 2).
4. Upstream transportation and distribution.
5. Waste generated in operations.
6. Business travel.
7. Employee commuting.
8. Upstream leased assets.

Downstream:

9. Downstream transportation and distribution.
10. Processing of sold products.
11. Use of sold products.
12. End-of-life treatment of sold products.
13. Downstream leased assets.
14. Franchises.
15. Investments.

In the Scope 3 GHG emissions inventory process, it is necessary to identify which of the 15 categories are relevant to the company's business (Table 6.5).

Table 6.5 Example of template for Scope 3 GHG emissions inventory

Emission source	Activity data	Units	Emission factor	EF units	Calculated emissions (tCO_2e)
Purchased goods and services		kg		kg CO_2e/ kg	
Upstream transportation and distribution		ton-km		kg CO_2e/ ton-km	
Waste generated in operations		kg		kg CO_2e/ kg	
Business travel		km		kg CO_2e/ km	

Source: Author elaborations

Ensuring Inventory Quality and Consistency

A robust Quality Management System (QMS) is fundamental to ensuring that a greenhouse gas (GHG) emissions inventory remains aligned with the core principles of the GHG Protocol Corporate Standard—namely, relevance, completeness, consistency, transparency, and accuracy. Moreover, a well-structured QMS enables organizations to proactively adapt to evolving regulatory requirements and stakeholder expectations in future GHG emissions programs.

The Key Steps for Implementing a QMS for GHG Inventories are: (i) Define Governance and Responsibilities to establish clear roles and responsibilities for data collection, calculation, verification, and reporting; (ii) Develop Standard Operating Procedures (SOPs), for data collection, emissions factor selection, calculation methodologies, and quality checks. The SOPs should be regularly reviewed and updated; (iii) Implement Data Quality Controls introducing automated and manual checks to validate data accuracy and completeness; (iv) Conduct Regular Internal Reviews scheduling periodic internal audits to assess compliance with the GHG Protocol and internal standards and collecting feedbacks from internal and external stakeholders; (v) Ensure Documentation and Transparency maintaining comprehensive records of data sources, assumptions, methodologies, and changes over time and, thus, facilitating third-party verification by ensuring traceability and transparency.

Accounting for GHG Reductions

Corporate GHG emissions reductions are assessed by comparing the company's current emissions inventory to a defined base year, which serves as a benchmark for measuring progress. This comparison enables organizations to quantify their decarbonization efforts over time and demonstrate accountability.

To ensure that these reductions are credible, science-aligned, and future-proof, companies are increasingly aligning their targets with the Science-Based Targets initiative (SBTi). The SBTi provides, in fact, a rigorous framework for setting emissions reduction goals that are consistent with the level of decarbonization required to limit global warming to well below 2 °C, and ideally to 1.5 °C, as outlined in the Paris Agreement (See Sect. 6.2.1 for details).

The GHG Protocol Corporate Standard employs a bottom-up methodology to calculate greenhouse gas emissions. This means emissions are first measured at the level of individual sources or facilities, and then aggregated to determine the company-wide total. This approach allows organizations to report emissions data at various levels of detail, such as by specific sources, individual facilities, or groups of facilities within a particular country and, at the same time, it allows the identification of hot spots (i.e. a source of emissions or category within the value chain that contributes significantly to the total GHG emissions) to drive specific emissions reduction initiatives.

GHG reduction targets (e.g., net-zero targets) could also be achieved from GHG mitigation projects that are external to the reporting entity's value chain and used to compensate, through the generation of offsets, for emissions that cannot be eliminated. Offsets are discrete GHG reductions used to compensate for GHG emissions and **the GHG Protocol allows companies to report offsets separately from their Scope 1, 2, and 3 emissions**. Offsets do not reduce reported emissions, but they can be used to claim carbon neutrality or net-zero status, provided they meet quality criteria (e.g., additionality, permanence, verification). Project reductions that are to be used as offsets should be quantified using a project quantification method that addresses the following accounting issues:

1. Selection of a baseline scenario and emission: The baseline scenario represents what would have happened in the absence of the project. Baseline emissions are the hypothetical emissions associated with this scenario.
2. Demonstration of additionality: This refers to whether the project has led to emission reductions or removals that go beyond what would have occurred without the project's implementation.
3. Identification and quantification of relevant secondary effects: Secondary effects are generally minor and unintended greenhouse gas (GHG) impacts resulting from a project. These include leakage as well as alterations in emissions occurring upstream or downstream of the project.
4. Consideration of reversibility: Certain mitigation projects achieve atmospheric CO_2 reductions through the capture, sequestration, and/or long-term storage of carbon or other greenhouse gases in biological (e.g., afforestation, reforestation, soil carbon enhancement) or non-biological (e.g., geological formations, mineralization) sinks. However, these carbon stocks are potentially non-permanent, as sequestered CO_2 may be re-emitted due to anthropogenic disturbances (e.g., biomass harvesting) or natural events (e.g., wildfires, pest outbreaks). Therefore, the risk of carbon reversal must be rigorously assessed, and the project design should incorporate appropriate permanence safeguards, including risk mitigation strategies, buffer reserves, or compensation mechanisms to address potential reversals.
5. Avoidance of double counting: To prevent double counting of emission reductions, offsets must originate from sources or sinks that lie outside the scope of the compliance target or emissions cap to which the offset is applied. Furthermore, in cases where the emission reductions occur at sources or sinks not directly owned or operated by the project participants (i.e., indirect reductions), it is essential to establish clear and verifiable ownership of the associated emission reductions. This ensures exclusive claim over the reductions and maintains the environmental integrity of the offset mechanism.

In the context of carbon credits markets (See Sect. 6.2.3 for details), offsets may be converted into carbon credits when applied toward compliance with externally imposed emissions reduction targets. Carbon credits are standardized, tradable instruments issued by accredited greenhouse gas programs, such as Verra (VCS), Gold Standard, or regulatory schemes like the EU ETS or California Cap-and-Trade. These credits represent verified reductions or removals of GHG emissions, typically quantified in metric tons of CO_2-equivalent.

For sustainability professionals, it is critical to understand that credits are not only transferable across entities but also subject to rigorous criteria for additionality, permanence, and third-party verification. Their issuance and retirement are tracked in registries to ensure transparency and prevent double counting. In both compliance and voluntary carbon markets, these instruments play a strategic role in enabling organizations to meet net-zero or carbon neutrality goals when direct emissions reductions are insufficient or infeasible.

Companies should disclose greenhouse gas (GHG) trades, such as the purchase or sale of offsets or credits, in their public GHG reports under the "optional information" section. This disclosure can be framed in relation to a specific emissions reduction target or the organization's corporate GHG inventory. To ensure transparency and maintain stakeholder confidence, the report should include relevant details that substantiate the credibility and integrity of the traded offsets or credits, such as the issuing program, verification status, and any associated risk mitigation measures.

Reporting GHG Emissions

The GHG Protocol Corporate Standard requires reporting a minimum of Scope 1 and Scope 2 emissions. A public GHG emissions report that is in accordance with the GHG Protocol Corporate Standard shall include the following information:

1. Description of the company and inventory boundary.
2. Information on emissions: Total scope, and separately, for Scope 1 and 2 emissions independent of any GHG trades such as sales, purchases, transfers, or banking of allowances; Emissions data for all six GHGs separately (CO_2, CH_4, N_2O, HFCs, PFCs, SF6) in metric tonnes and in tonnes of CO_2 equivalent; Year chosen as base year, and an emissions profile over time; appropriate context for any significant emissions changes that trigger base year emissions recalculation; emissions data for direct CO_2 emissions from biologically sequestered carbon; methodologies used to calculate or measure emissions.

Among optional Information a public GHG emissions report could include, there are emissions data from relevant Scope 3 emissions activities, a description of performance measured against internal and external benchmarks, relevant ratio performance indicators; information on offsets, an outline of any external assurance provided, and a copy of any verification statement.

Relative to the GHG verification process, the engagement of a verifier can occur at various points during the GHG preparation and reporting process. Some companies may establish a semi-permanent internal verification team to ensure that GHG data standards are being met and improved on an ongoing basis.

Setting a GHG Target

A sound business strategy relies on setting measurable targets for key performance indicators such as revenue, sales, and operational efficiency, and systematically tracking progress against them. Similarly, effective greenhouse gas (GHG) management requires the establishment of clear GHG reduction targets, supported by ongoing performance monitoring and evaluation.

Common drivers for setting a GHG target include:

- Minimizing and managing GHG Risks;
- Achieving costs saving and fostering innovation;
- Preparing for future regulations;
- Demonstrate to stakeholders corporate responsibility;
- Participating in voluntary programs.

There are two types of target:

- **Absolute**: Usually expressed in terms of a reduction over time in a specified quantity of GHG emissions (e.g., 1% reduction each year).
- **Intensity-based**: Usually expressed as a reduction in the ratio of GHG emissions relative to another business metric (e.g., reduce GHGs by 1% per production unit).

The steps to define a GHG Target are the following (Fig. 6.4):

Fig. 6.4 10 Steps to define a GHG Target. Source: Author elaborations on the basis of GHG Protocol

It is worth to underline that the GHG Protocol is actively updating its standards to align with evolving regulatory frameworks such as the CSRD, ISSB, and SEC climate disclosure rules.[12]

A technical working group is currently reviewing and refining aspects of Scope 1 reporting as part of a broader update to the Corporate Standard. These updates are expected to address:

- Greater comparability of GHG data;
- Clarification of consolidation approaches;
- Enhanced alignment with mandatory disclosure requirements.

It is worth to underline that GHG emissions reporting is a critical component of corporate climate governance, serving not merely as a regulatory compliance requirement but also as a strategic enabler for internal carbon pricing mechanisms and effective engagement in carbon credit markets.

Regarding the Internal Carbon Pricing, GHG inventories allow companies to assign a monetary value to each ton of CO_2e emitted, facilitating the implementation of internal carbon pricing models. These models incentivize low-carbon decision-making, support capital allocation toward decarbonization technologies, and enhance climate risk management in line with frameworks like the Standard-Based Targets Initiative, SBTi.

6.2.1 SBTi: Science-Based Targets Initiative

The Science-Based Targets initiative (SBTi) is a corporate climate action organization that enables companies and financial institutions worldwide to play their part in combating the climate crisis by setting science-based targets for reducing greenhouse gas emissions. These targets align with the goals of the Paris Agreement, aiming to limit global warming to well below 2 °C, ideally to 1.5 °C.

The Science-Based Targets initiative (SBTi) defines several types of targets to help companies reduce their greenhouse gas (GHG) emissions. Here are the main types of targets:

- **Near-Term Targets:** These targets focus on immediate actions that companies can take to reduce their emissions within the next 5–10 years. They are designed to ensure that companies make significant progress toward reducing their carbon footprint in the short term.
- **Long-Term Targets:** Long-term targets are set for a period extending beyond 10 years, aiming to achieve substantial reductions in emissions by mid-century. These targets are aligned with the goal of reaching net-zero emissions by 2050.

[12] GHG Protocol (2025). Overview of GHG Protocol Integration in Mandatory Climate Disclosure Rules.

- **Net-Zero Targets:** Net-zero targets involve reducing emissions to as close to zero as possible and balancing any remaining emissions with an equivalent amount of carbon removal. Companies are encouraged to set net-zero targets that are consistent with limiting global warming to 1.5 °C above pre-industrial levels.
- **Sector-Specific Targets:** SBTi provides tailored guidance for different sectors, recognizing that industries face unique challenges in reducing emissions. This includes specific pathways for high-emitting sectors, which are crucial for achieving global climate goals.

The targets (near-term and long–/net-zero term) shall cover at least 95% of company-wide Scope 1 and Scope 2 emissions.

Scope 3 emission, instead, shall be included in Near-Term targets if they represent 40% or more of the total GHG emissions. In this case, the target boundary must cover minimum 67% of Scope 3 emissions. The boundary of Scope 3 emissions for long-term science-based targets shall cover at least 90% of total Scope 3 emissions according to GHG Protocol (See: Bhatia et al., 2011)[13,,14]

Focus

Summary on Scope 1, 2, and 3 Emissions Targets

Scope 1: Direct emissions from owned or controlled sources.

Scope 2: Indirect emissions from the generation of purchased electricity, steam, heating, and cooling consumed by the reporting company. Companies shall disclose whether they are using a location- or market-based accounting approach as per the GHG Protocol Scope 2 Guidance[15] to calculate base year emissions and to track performance against a science-based target.

Scope 3: All other indirect emissions that occur in a company's value chain. At a minimum, near-term Scope 3 targets (covering total required Scope 3 emissions or individual Scope 3 categories) shall be aligned with methods consistent with the level of decarbonization required to keep global temperature increase well below 2 °C compared to pre-industrial temperatures. For long-term Scope 3 targets, this minimum ambition is increased to 1.5 °C.

SBTi Corporate Net-Zero Standard Criteria[16] requires companies, as general rules, to define and submit targets only at the parent- or group level, not the

[13] Bhatia P., C. Cummis, D. Rich, L. Draucker, H. Lahd and A. Brown (2011). Corporate value chain (Scope 3) accounting and reporting standard. https://ghgprotocol.org

[14] Science Based Targets. (2024b). Getting started guide. Retrieved from https://sciencebased-targets.org

[15] World Resources Institute & World Business Council for Sustainable Development. (2023). Scope 2 guidance. https://ghgprotocol.org

[16] Science Based Target (2024a). Corporate Net zero Target Criteria, version 1.2, March 2024.

subsidiary level.[17] Target to be defined is relative to Greenhouse Gas (GHG) Emissions reduction; therefore, they shall cover all relevant emissions of the seven GHGs[18] required by the GHG Protocol Corporate Standard.[19] It is worth to underline that Financial Institutions must follow a separate Science-Based Standard Criteria. In particular, since July 2025, for the first time, banks, asset owners and managers, private equity firms, and other financial institutions have been able to set science-based targets aligned with net-zero. The new Standard provides clear, actionable guidance to align lending, investment, insurance, and capital markets activities with limiting global warming and achieving net-zero by 2050 at the latest.[20]

The companies shall publicly set a science-based target that clearly and transparently communicates each of the target's relevant components including:

- The target year;
- Magnitude of emissions reductions that will be achieved for near-term and long-term science-based targets;
- A base year.

Targets must be modeled using the latest version of methods and tools[21] approved by the SBTi.

SBTi provides two different tools[22]:

- **SBTi Corporate Near-Term Tool**: The SBTi Corporate Near-Term tool uses data from International Energy Agency (IEA) Publications, in particular for the definition of 1.5 °C aligned pathways (2021).[23]
- **Corporate Net-Zero Tool**: The tool uses data from International Energy Agency (IEA) (2021) and also SBTi (2021).[24]

The tools are provided in excel files, even for specific sectors, and include an input-data table where the companies must fill data about the base year level of Scope 1, Scope 2, and Scope 3 emissions and the target year for emission reduction. The tools automatically provide the Scope 1 and 2 and Scope 3 science-based emission reduction targets, i.e. targets aligned to the goals of the Paris Agreement, aiming to limit global warming to well below 2 °C, ideally to 1.5 °C.

---

[17] Parent companies shall include the emissions of all subsidiaries in their target submission.

[18] The seven GHGs are carbon dioxide (CO2), methane (CH4), nitrous oxide (N2O), hydrofluorocarbons (HFCs), perfluorocarbons (PFCs), sulfur hexafluoride (SF6), and nitrogen trifluoride (NF3).

[19] Greenhouse Gas Protocol. (2015). GHG protocol revised standards. Retrieved from https://ghgprotocol.org

[20] https://sciencebasedtargets.org/financial-institutions

[21] https://sciencebasedtargets.org/resources/

[22] https://sciencebasedtargets.org/resources?tab=technical-development#resource

[23] International Energy Agency (IEA). (2021). Net Zero by 2050: A roadmap for the global energy sector. IEA. https://www.iea.org/reports/net-zero-by-2050

[24] Science Based Targets initiative (SBTi). (2021). Foundations for science-based net-zero target setting in the corporate sector. Science Based Targets initiative.

After Target commitment, Target validation is carried out by SBTi Services, a wholly-owned subsidiary of the Science-Based Targets initiative. SBTi Services independently assesses corporate and financial institutions' science-based emission reduction targets against SBTi Standards. Criteria Assessment Tables are available in the website[25] and they outline the major checks conducted by SBTi Services during the validation process, to ensure conformance with SBTi Criteria.

Some of the indicators considered as relevant (not exhaustive list is reported) to assess and validate near-term targets are relative to:

- Organizational boundary: Companies should submit targets only at the parent- or group level, not the subsidiary level. Parent companies shall include the emissions of all subsidiaries in their target submission, in accordance with the boundary criteria.
- Greenhouse gases: The targets shall cover all relevant emissions according to the requirement of the GHG Protocol.
- Scope 1 and Scope 2: The targets shall cover company-wide Scope 1 and Scope 2 emissions, as defined by the GHG Protocol.
- Scope 3: If a company's relevant Scope 3 emissions are 40% or more of total Scope 1, 2, and 3 emissions, they shall be included in near-term science-based targets.
- Scope 1, 2, and 3 allowable exclusions: Companies shall not exclude more than 5% of total combined Scope 1 and Scope 2 emissions from either the boundary of the GHG inventory or the target boundary.
- Scope 3 emissions coverage for near-term targets: Companies shall set one or more emission reduction near-term targets and/or supplier or customer engagement targets that collectively cover at least 67% of total reported and excluded Scope 3 emissions considering the minimum boundary of each category in conformance with the GHG Protocol Corporate Value Chain (Scope 3) Accounting and Reporting Standard.
- Method validity: Targets must be modeled using the latest version of methods and tools approved by the SBTi.
- Scope 2 accounting approach: Companies shall disclose whether they are using a location- or market-based accounting approach, as per the GHG Protocol Scope 2 Guidance, to calculate base year emissions and to track performance against a science-based target.
- Scope 3 inventory: Companies shall complete a Scope 3 inventory covering gross Scope 3 emissions for all its relevant emissions sources according to the GHG Protocol Corporate Value Chain (Scope 3) Accounting and Reporting Standard.
- Carbon credits: The use of carbon credits must not be counted as emission reductions toward the progress of companies' near-term or long-term science-based targets (See Sect. 6.2.3 for details on the Carbon Credit Market).

[25] https://docs.sbtiservices.com/resources/CriteriaAssessmentIndicators.pdf

- Base and target years: Absolute and intensity-based emission reduction near-term targets must cover a minimum of 5 years and a maximum of 10 years from the date the target is submitted to the SBTi for validation.
- Progress to date: The minimum forward-looking ambition of near-term targets covering Scope 1 and/or Scope 2 emissions is consistent with reaching net-zero by 2050 at the latest

In case of net-zero targets validation, the indicators considered as relevant are in part similar to the above reported, but they also regard, among other:

- Scope 3 emissions coverage for long-term targets: The boundary of long-term science-based targets shall cover at least 90% of total Scope 3 emissions considering the minimum boundary of each category in conformance with the GHG Protocol Corporate Value Chain (Scope 3) Accounting and Reporting Standard.
- Carbon credits: The use of carbon credits shall not be counted as emission reductions toward the progress of companies' near-term or long-term science-based targets. Carbon credits may only be considered as an option for neutralizing residual emissions.
- State of net-zero emissions: Companies shall set one or more targets to reach a state of net-zero emissions, which involves: (a) reducing Scope 1, 2, and 3 emissions to zero or a residual level consistent with reaching net-zero emissions at the global or sector level in eligible 1.5 °C scenarios or sector pathways and (b) neutralizing any residual emissions at the net-zero target date—and any GHG emissions released into the atmosphere thereafter.
- Base year: The base year shall be no earlier than 2015. The company shall use the same base year for its long-term science-based targets as its near-term science-based targets. Scope 1 and Scope 2 targets shall use the same base year.
- Long-term targets shall have a target year no later than 2050.
- Level of ambition for Scope 1 and 2 targets: At a minimum, Scope 1 and Scope 2 targets shall be consistent with the level of decarbonization required to keep global temperature increase to 1.5 °C compared to pre-industrial temperatures.
- Level of ambition for Scope 3 emissions reductions targets: At a minimum, near-term Scope 3 targets shall be aligned with methods consistent with the level of decarbonization required to keep global temperature increase well below 2 °C compared to pre-industrial temperatures.

It is important to underline that Companies shall publicly report information pertaining to progress against validated targets, including separately reporting emissions and removals in the annual GHG inventory. In particular: (i) Companies must report annually their gross GHG emissions; (ii) Companies must report annually their carbon removals separately from their gross GHG emissions; and (iii) Companies must report annually their progress toward validated targets, e.g. amount of Scope 1, Scope 2, and Scope 3 emissions reduced in the reporting year.

Focus

How Many Companies Use Science-Based Targets? The SBTi Dashboard

As per April 2025, about 10,330 companies are committed to reach science-based targets. About 7502 of them have had their target validated by SBTi Services, a wholly-owned subsidiary of the Science-Based Targets initiative.

More than one half of companies with validated Science-based target are located in Europe, followed by Asia (33%) and Northern America (12%), See Table 6.6.

The first sector in terms of number of companies with science-based validated target is the one related to Professional Services, followed by Electrical Equipment and Machinery and Software and Services (See Fig. 6.5).

Table 6.6 Number of companies with validated target by region

Region	N of companies with validated target	%
Europe	3813	51%
Asia	2511	33%
Northern America	868	12%
Latin America and the Caribbean	136	2%
Oceania	107	1%
Africa	45	1%
MENA	22	0%
World	**7502**	

Source: Author elaboration on the basis of the SBTi Dashboard, by company excel file (Updated on the fourth of April 2025)

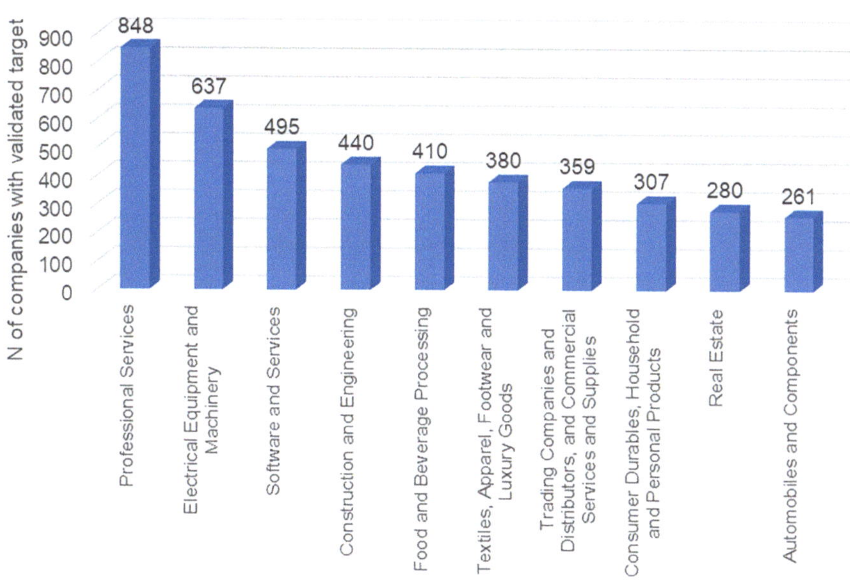

Fig. 6.5 Number of companies with validated target by TOP 10 Sectors. Source: Author elaboration on the basis of the SBTi Dashboard, by company excel file (Updated on the fourth of April 2025)

6.2.2 MACC, A Tool to Define Decarbonization Strategies

Strategic consultants and sustainability professionals must help clients in navigating the GHG scopes by identifying high-impact areas (hot spots) and prioritizing interventions to drive an effective decarbonization process. Tools such as marginal abatement cost curves (MACCs) are instrumental in this phase.

The Marginal Abatement Cost Curve (MACC) is a strategic decision-support tool used to evaluate and prioritize greenhouse gas (GHG) reduction measures based on their cost-effectiveness and emissions reduction potential.

Each bar on the curve represents a specific mitigation option, showing:

- Width: The emissions reduction potential of a specific mitigation measure (typically in tonnes of CO_2e per year). In Fig. 6.6, it is reported an example with the following measures: energy-efficiency upgrades, renewable energy deployment, electrification, and process optimization.
- Height: The marginal cost of abatement (typically in €/tCO_2e) measured as the ratio between the net cost of the measure and the total emission reduced.

The curve is typically sorted from the lowest-cost (or even cost-saving) measures on the left to the highest-cost measures on the right. The X-axes report the Cumulative width, i.e. it shows the total abatement potential across all measures.

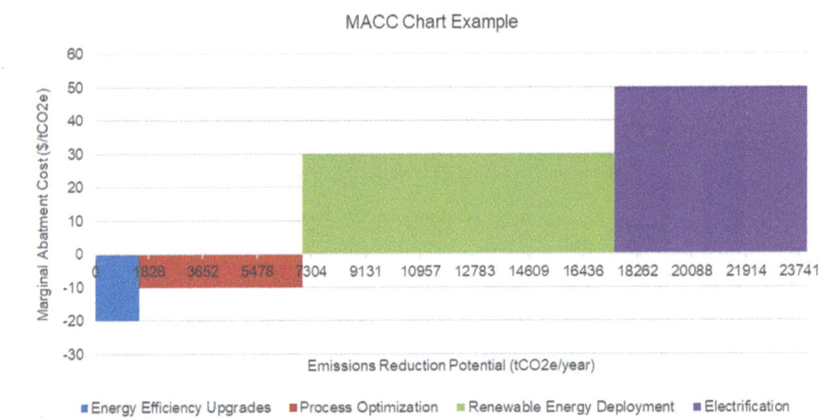

Mitigation Measure	Emissions Reduction Potential (tCO2e/year)	Marginal Abatement Cost (€/tCO2e)
Energy Efficiency Upgrades	1500	-20
Process Optimization	5500	-10
Renewable Energy Deployment	10500	30
Electrification	6500	50

Fig. 6.6 Example of MACC Chart. Source: Author Elaboration

This type of visualization helps sustainability professionals prioritize actions based on both economic efficiency and climate impact. The curve helps identify which measures are the most cost-effective and which become increasingly expensive and helps, thus, the company sequence investments, starting with low-cost, high-impact measures while planning for more capital-intensive solutions later.

Another way to interpret the MACC is through the lens of investment: it shows the carbon price (e.g., in USD per ton of CO_2) at which a particular emission reduction measure becomes financially sustainable—meaning the savings from avoided CO_2 emissions just balance out the cost of implementing the measure (i.e., the net present value (NPV) is zero).

In this way, MACC becomes a strategic lens through which organizations can view their decarbonization journey. By combining economic and environmental insights, the MACC empowers decision-makers to act with clarity, confidence, and impact.

Focus

The Shipping Industry: A Case Study of MACC Application

The maritime sector plays a critical role in the global economy, facilitating the movement of goods and people across the world. However, it is also a significant contributor to greenhouse gas (GHG) emissions, accounting for approximately 3% of global GHG emissions. The International Maritime Organization (IMO) in its report "*Fourth IMO GHG Study 2020*" stated, in fact, that in 2018, total shipping (including international, domestic, and fishing) emitted approximately 1076 million tonnes of GHGs, of which 1056 million tonnes were CO_2. This accounted for about 2.89% of global anthropogenic CO_2 emissions.

Recognizing the need for urgent action to mitigate climate change, both the International Maritime Organization (IMO) and the European Union (EU) have implemented comprehensive frameworks to reduce GHG emissions from international shipping.

The International Maritime Organization, in particular, has stated the ambition to achieve net-zero GHG emissions by 2050, as well as indicative checkpoints to reach 20%, striving for 30%, total GHG emission reduction in 2030, and 70%, striving for 80%, reduction in 2040.

A wide range of abatement measures can be deployed in the shipping fleet, from hull and machinery improvements, operational measures and speed reductions, to alternative carbon-neutral fuels like biodiesel, hydrogen and ammonia, as well as onboard carbon capture and subsequent storage.

The Longva et al. (2024) Study "Marginal abatement cost curves for CO_2 emission reduction from shipping to 2050"[26] has modeled the CO2 emission

[26] Longva, T., Eide, M. S., Endresen, Ø., Sekkesæter, Ø., Helgesen, H., & Rivedal, N. H. (2024). Marginal abatement cost curves for CO2 emission reduction from shipping to 2050. Maritime Transport Research, 6.

abatement potentials and costs toward 2050 for all the ships above 400 gross tons using the MACC tool approach. The study incorporates over 50 energy-efficiency measures, 10 fuel systems, and 8 alternative fuels to build global MACCs.

They developed an abatement evaluation module which calculates the Net Present Value (NPV) and CO_2 reduction of abatement measures for each ship in the fleet and applies the compliant measure with highest NPV. Since, they found that applying different speed reduction levels, fuel systems and energy-efficiency measures, the combined effect may not be the sum of the isolated effects of individual measures, to take into account the interaction between measures, they considered these measures in packages and groups.

Each measure group (mg) has a *NPV* and CO_2 reduction potential (*RED*) over the investment horizon. *NPV* and *RED* are relative to a basic energy-efficiency package, no speed reduction, and a conventional VLSFO/MGO diesel engine. The marginal abatement cost (*MAC*), in USD per ton of CO_2 reduced for each group of measures, is calculated according to the Ratio NPV/RED starting from the *mg* with the highest *NPV* that is added to the MACC as the first *mg*.

For all other measure groups with a RED higher than the first added group, they have recalculated MAC_{mg} relative to the most recently added group as $\Delta NPV/\Delta RED$. The group with the lowest *MAC* is added to the MACC as the next.

The results of the study show that a 20–30% reduction in emissions by 2030 is achievable at a marginal cost of $50–100 per ton of CO2, a 70–80% reduction by 2040 would cost $230–240 per ton of CO_2, and the net-zero emissions by 2050 could be reached at $300 per ton of CO_2.

6.2.3 The Carbon Credit Markets

The concept of carbon trading emerged from the Kyoto Protocol (1997),[27] which introduced market-based mechanisms like the Clean Development Mechanism (CDM) and Joint Implementation (JI) to help countries meet their emission reduction targets.[28]

[27] Kyoto Protocol to the United Nations Framework Convention on Climate Change. (1997). United Nations Treaty Series, Vol. 2303, p. 162.

[28] The CDM allows industrialized countries to finance emission-reduction projects in developing countries and use the resulting credits to meet their own emission targets. Joint Implementation (JI) allows a country with a Kyoto emission-reduction commitment (Annex B Party) to implement an emission-reduction or carbon-removal project in another Annex B country and earn Emission Reduction Units (ERUs), each equal to 1 tonne of CO2 equivalent.

In 2005, the European Union launched the EU Emissions Trading System (EU ETS), the world's first and largest international carbon market. It operates on a cap-and-trade principle, where a cap is set on total emissions and allowances are traded among regulated entities. However, the number of emissions trading systems around the world is increasing, in fact national or sub-national systems similar to EU ETS are already operating or under development in Canada, China, Japan, New Zealand, South Korea, Switzerland, and the United States.[29] Today there are 80 carbon taxes and emissions trading systems (ETSs) in operation worldwide.[30]

Carbon taxes and emissions trading systems are, thus, designed to reduce global greenhouse gas (GHG) emissions by putting a price on carbon. They operate by allowing entities, governments, or companies, to buy and sell carbon credits, each representing the right to emit one metric ton of carbon dioxide (CO_2) or its equivalent in other GHGs. This market-based approach incentivizes emission reductions by making pollution financially accountable, thereby channeling capital toward climate mitigation projects such as reforestation, renewable energy, or carbon capture.[31]

In order to provide a picture of the Carbon Credit Market Scale, World Bank in its recent Report "State and trends of carbon pricing 2025" states that, in 2024, approximately 290 million carbon credits were issued by major independent crediting mechanisms. The global pool of unretired credits from these mechanisms (as a proxy of supply) approached 1 billion tons. Retirements, considered as a proxy for demand, rose by 15%, driven primarily by compliance markets, which accounted for nearly 25% of all retirements. Voluntary buyers shifted toward nature-based removals and clean cooking projects, with retirements in these categories increasing by 25% and 50%, respectively. Renewable energy credits remained significant but saw a slight decline in retirements (Fig. 6.7).

The **compliance carbon markets** are regulatory systems where governments require certain entities, usually large emitters like power plants, industrial facilities, or fuel suppliers, to pay for their greenhouse gas (GHG) emissions. They are regulated by governments or international bodies and are designed to help countries or sectors meet legally binding emissions reduction targets. Participants in these markets must surrender allowances or credits to cover their emissions, with penalties for non-compliance.

Key examples of carbon markets include:

- The EU Emissions Trading System (EU ETS) which is the largest and most established compliance market.
- The California Cap-and-Trade Program: a state-level market in the U.S.

---

[29] https://climate.ec.europa.eu/eu-action/carbon-markets/eu-emissions-trading-system-eu-ets/international-carbon-market_en

[30] World Bank. (2025). State and trends of carbon pricing 2025. World Bank. https://doi.org/10.1596/978-1-4648-2255-1

[31] World Bank. (2025). State and trends of carbon pricing 2025. World Bank. https://doi.org/10.1596/978-1-4648-2255-1

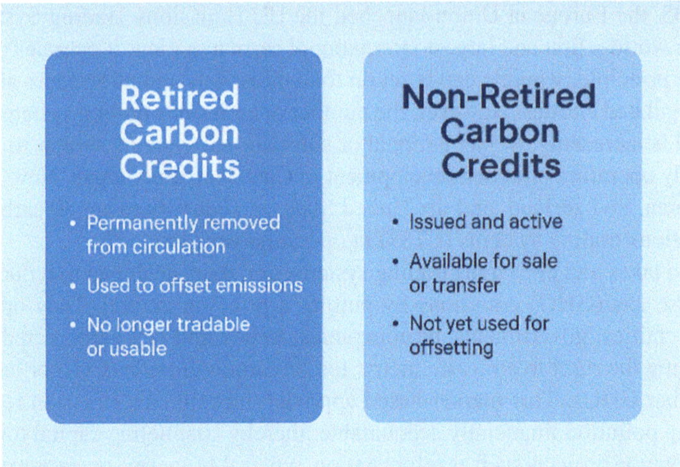

Fig. 6.7 The carbon credit market: differences between the concepts of retired and non-retired carbon credits. Source: Author Elaboration

- The Korea ETS and China's National ETS: emerging large-scale systems in Asia.
- The Clean Development Mechanism (CDM): It has been established under the Kyoto Protocol, allowing emission reduction projects in developing countries to earn certified emission reduction (CER) credits.

Voluntary carbon markets (VCMs) enable non-state actors like corporations to purchase carbon credits on a discretionary basis to compensate for their greenhouse gas (GHG) emissions. These credits, typically issued by independent crediting mechanisms such as Verra or Gold Standard, represent verified emissions reductions or removals and are often used to support corporate climate strategies, including net-zero targets and broader environmental, social, and governance (ESG) commitments (Climate Focus, 2025[32]; ICVCM, 2024[33]).

VCMs differ from compliance markets since participation is not mandated by regulation, but rather driven by reputational, strategic, or pre-emptive regulatory motivations. The market has seen increasing demand for credits associated with high environmental integrity, particularly nature-based removals and clean cooking technologies, reflecting evolving buyer preferences and the growing role of carbon markets in private sector decarbonization (World Bank, 2025, pp. 46–50).

[32] Climate Focus. (2025). The Voluntary Carbon Market Dashboard. https://climatefocus.com/initiatives/voluntary-carbon-market-dashboard/

[33] Integrity Council for the Voluntary Carbon Market (ICVCM). (2024). Core Carbon Principles and Assessment Framework. https://icvcm.org

Key features thus include:

- No legal obligation to participate.
- Credits are typically Verified Emission Reductions (VERs).
- Projects include reforestation, renewable energy, methane capture, clean cooking, and more.
- Standards include Verra's Verified Carbon Standard (VCS),[34] Gold Standard, and Climate Action Reserve.

Major corporates like Microsoft, ENI, and Google use VCMs to meet climate goals beyond regulatory requirements.

FOCUS: ENI S.p.A. Clean Cooking Projects

Eni launched its Clean Cooking program in 2018, aiming to promote the replacement of traditional stoves with more efficient models in sub-Saharan Africa. The program seeks to reduce wood and charcoal consumption by over 60%, improving the quality of life for participating families and lowering the carbon footprint associated with cooking activities (Source: ENI website).

In 2024, Eni reaffirmed its commitment during the Clean Cooking Summit for Africa, organized by the International Energy Agency (IEA), with the goal of distributing improved stoves to ten million people in sub-Saharan Africa by 2027, and reaching 20 million people with advanced cooking solutions by 2030.

The Clean Cooking program has had a significant impact on the company, contributing to the reduction of greenhouse gas emissions and generating carbon credits used in alignment with Eni's Net-Zero strategy. Additionally, the program has supported local entrepreneurship and economic activities within communities by encouraging local production of cooking systems and facilitating access to technology, capital, and markets.

Eni has also signed the "Clean Cooking Declaration: Making 2024 the pivotal year for Clean Cooking",[35] to accelerate universal access to modern cooking systems—essential for ensuring affordable, reliable, and sustainable energy for all, as outlined in the United Nations Sustainable Development Goal 7.

In conclusion, the participation in the carbon credit market can actively foster decarbonization by creating financial incentives, enabling strategic investment, and accelerating the adoption of low-carbon technologies. Entities in fact are

[34] Verra registry's methodology in accordance with the sectors and project activities. See: https://verra.org/program-methodology/vcs-program-standard/overview/

[35] International Energy Agency (IEA). (2024, May 14). Clean Cooking Declaration: Making 2024 the pivotal year for clean cooking. https://www.iea.org/news/the-clean-cooking-declaration-making-2024-the-pivotal-year-for-clean-cooking

incentivated to either: (i) reduce their own emissions to generate tradable credits, or (ii) purchase credits to offset residual emissions.

Carbon credit purchases can be part of a transitional strategy, allowing companies to offset emissions while investing in long-term decarbonization.

References

Beloin-Saint-Pierre, D., Albers, A., Hélias, A., Tiruta-Barna, L., Fantke, P., Levasseur, A., Benetto, E., Benoist, A., & Collet, P. (2020). Addressing temporal considerations in life cycle assessment. Science of the Total Environment, 743, Article 140700. https://doi.org/10.1016/j.scitotenv.2020.140700 [backend.orbit.dtu.dk]

Bhatia, P., Cummis, C., Rich, D., Draucker, L., Lahd, H., & Brown, A. (2011). *Corporate value chain (scope 3) accounting and reporting standard.* https://ghgprotocol.org

Crippa, M., Guizzardi, D., Pagani, F., Banja, M., Muntean, M., Schaaf, E., Monforti-Ferrario, F., Becker, W. E., Quadrelli, R., Risquez Martin, A., Taghavi-Moharamli, P., Köykkä, J., Grassi, G., Rossi, S., Melo, J., Oom, D., Branco, A., San-Miguel, J., Manca, G., Pisoni, E., Vignati, E., & Pekar, F. (2024). *GHG emissions of all world countries.* Publications Office of the European Union.

European Commission - Joint Research Centre - Institute for Environment and Sustainability. (2010). *International Reference Life Cycle Data System (ILCD) Handbook: General guide for Life Cycle Assessment - Detailed guidance* (1st ed., EUR 24708 EN). Publications Office of the European Union.

Finnveden, G., Hauschild, M. Z., Ekvall, T., Guinée, J., Heijungs, R., Hellweg, S., Köhler, A., Pennington, D., & Suh, S. (2009). Recent developments in life cycle assessment. *Journal of Environmental Management, 91*(1), 1–21. https://doi.org/10.1016/j.jenvman.2009.06.018

GHG Protocol. (2025). *Overview of GHG protocol integration in mandatory climate disclosure rules.*

Greenhouse Gas Protocol. (2015). *GHG protocol revised standards.* https://ghgprotocol.org

International Energy Agency (IEA). (2021). *Net zero by 2050: A roadmap for the global energy sector.* IEA. https://www.iea.org/reports/net-zero-by-2050

International Energy Agency (IEA). (2024, May 14). *Clean cooking declaration: Making 2024 the pivotal year for clean cooking.* https://www.iea.org/news/the-clean-cooking-declaration-making-2024-the-pivotal-year-for-clean-cooking

International Sustainability Standards Board (ISSB). (2023). *IFRS S2 climate-related disclosures.*

IPCC. (1996). *Revised 1996 IPCC guidelines for national greenhouse gas inventories.*

Kyoto Protocol to the United Nations Framework Convention on Climate Change. (1997). United Nations Treaty Series, Vol. 2303, p. 162.

Longva, T., Eide, M. S., Endresen, Ø., Sekkesæter, Ø., Helgesen, H., & Rivedal, N. H. (2024). Marginal abatement cost curves for CO_2 emission reduction from shipping to 2050. *Maritime Transport Research, 6.*

Science Based Target initiative (SBTi). (2024a, March). *Corporate net zero target criteria, version 1.2.*

Science Based Target initiative (SBTi). (2024b). *Getting started guide.* https://sciencebasedtargets.org

Science Based Targets initiative (SBTi). (2021). *Foundations for science-based net-zero target setting in the corporate sector.* Science Based Targets initiative.

World Bank. (2025). *State and trends of carbon pricing 2025.* World Bank. https://doi.org/10.1596/978-1-4648-2255-1

World Resources Institute & World Business Council for Sustainable Development. (2004). *The greenhouse gas protocol: A corporate accounting and reporting standard* (Revised ed.).

World Resources Institute & World Business Council for Sustainable Development. (2023). *Scope 2 guidance.* https://ghgprotocol.org

7 Integrating Sustainable Practices in Business

Although companies are investing heavily in environmentally and socially responsible products, many see only limited financial returns and face difficulties turning their sustainability efforts into profitable ventures. As a result, business leaders around the world are increasingly frustrated, finding it hard to effectively integrate sustainability into their core business strategies.

This Chapter shows how the concepts learned in the previous chapters can be integrated into the corporate decision-making process to define strategic plans and maximize profits. Integrating sustainability into business is, in fact, not only an ethical commitment but also a strategic imperative that enhances risk management, stakeholder communication, and long-term corporate value.

Recent literature emphasizes that embedding sustainability into corporate strategy requires a multidimensional approach, balancing economic, environmental, and social objectives across all organizational levels.[1] This integration supports more resilient and adaptive business models, especially in the face of increasing regulatory pressures and stakeholder expectations.[2] Moreover, sustainability-oriented strategies have been shown to foster innovation, improve brand reputation, and open new market opportunities, ultimately contributing to competitive advantage and profitability.[3]

[1] Nguyen, H. L., & Kanbach, D. K. (2023). Toward a view of integrating corporate sustainability into strategy: A systematic literature review. Corporate Social Responsibility and Environmental Management.

[2] Oertwig, N., et al. (2017). Integration of Sustainability into the Corporate Strategy. In Sustainable Manufacturing. Springer.

[3] Johnson, E., C. Ramirez and A. Oliveira (2024) Integrating Sustainability into Corporate Strategy: Challenges and Opportunities. International Journal of Management, Business, and Economics.

M. Gussoni, *Integrating Strategic Sustainability Practices in Business*, CSR, Sustainability, Ethics & Governance,
https://doi.org/10.1007/978-3-032-09617-3_7

Thakkar (2025)[4] shows how sustainability can be embedded into core business strategies, exploring its impact on long-term value creation and competitive advantage. Among best practices, it lists setting measurable sustainability goals, fostering cross-functional collaboration, and leveraging innovation to align business objectives with ESG priorities.

Also business model innovation has recently gained momentum in both academic research and practical application (Evans et al., 2017).[5] Business Model innovation, when embedding sustainability principles, is increasingly recognized as a critical strategy for fostering sustainability-oriented innovation. Despite this growing interest, however, there remains limited understanding of how sustainable business models (SBMs) can be successfully implemented. This chapter aims to propose a comprehensive framework that enhances the understanding of business model innovations capable of improving an organization's economic, environmental, and social performance. By doing so, it seeks to bridge the gap between conceptual development and practical application, offering guidance for organizations striving to integrate sustainability into their core business logic.

7.1 Circular and Sustainable Business Models (SBMs) Design

According to Spieth et al. (2014), the business model concept extends well beyond a simple narrative of how a company operates, it offers a comprehensive view of the firm's activities and strategy. Business models refer to the logic of how a firm does business, and explain how the firm creates, delivers and captures value. Teece (2010) emphasizes that business model design facilitates the reconfiguration of organizational capabilities, enabling firms to adapt to evolving market conditions. Business models are, thus, increasingly regarded not only as tools for innovation and commercialization of technological advancements but also as innovation subjects themselves (Teece, 2010; Zott et al., 2011).

Given these premises, a business modeling consulting service supports organizations in creating, refining, and innovating their business by aligning strategic goals with market dynamics, customer needs, and sustainability imperatives. In particular, clients often require services to design new or improved business models that leverage innovation and sustainability.

Often, in strategic consulting, Business Models innovation is defined according to a Business Model Canvas Framework.

The Business Model Canvas, developed by Alexander Osterwalder,[6] is a strategic management tool that helps visualize and describe how a business creates, delivers, and captures value. It consists of nine building blocks to be filled answering to specific questions:

[4] Thakkar, R. (2025) Integrating Sustainability into Corporate Strategy: Challenges and Best Practices. Vol. 3 No. 03 (2025): IRJAEM Vol.03 Issue 03- [MARCH 2025].

[5] Evans, S., D. Vladimirova, M. Holgado, K.V. Fossen, M. Yang, E.A. Silva and C.Y. Barlow (2017) Business Model Innovation for Sustainability: Towards a Unified Perspective for Creation of Sustainable Business Models. Business Strategy and the Environment, 26, 597–608 (2017).

[6] Osterwalder, A., & Pigneur, Y. (2010). Business model generation. John Wiley & Sons.

1. **Value Propositions**—What value do you deliver to the customer? The role of the strategic consultant is here to help the client in defining the competitive advantage of the product or the service to be introduced in the market. Michael E. Porter, a leading authority on competitive strategy, defines competitive advantage as "The ability of a firm to outperform its rivals by delivering greater value to customers or by achieving lower costs."[7] Porter outlines that competitive advantage arises from a firm's ability to either: (i) differentiate its products or services in ways that are valued by customers (Differentiation Strategy), or (ii) achieve cost leadership by producing goods or services at a lower cost than competitors (Cost Leadership Strategy). It is worth to underline however, that also a good sustainability strategy could drive to a competitive advantage.[8] As outlined in previous Chapters, it has been frequently demonstrated a positive correlation between ESG Risks management and financial performance. Michael Porter himself has acknowledged the strategic role of sustainability in his concept of Creating Shared Value,[9] which argues that companies can enhance their competitiveness while simultaneously advancing social and environmental conditions in the communities where they operate.
2. **Customer Segments**—Who are your target customers? Here the strategic consultant usually provides Market Research services for market segmentation and definition of strategies for pricing and market uptake. The market research should also provide forecasting for operating costs and sales that will be the base for business planning.
3. **Channels**—How do you reach your customers? This box should be filled with the ways the company communicates with and reaches its customers to deliver the product or service. Hence, it deals with marketing and communication services.
4. **Customer Relationships**—How do you interact with customers? This Box defines the type of relationship a company establishes with each Customer Segment. It reflects how it acquires, retains, and grows its customer base. This part deals with commercial and customer support services.
5. **Revenue Streams**—How does the business earn money? All the elements above should enable revenue streams which will be included in the project future business plan. The Types of Revenue Streams could be: (i) Transaction-based, i.e. one-time payments (e.g., product sales, event tickets); (ii) Recurring, i.e. ongoing payments (e.g., subscriptions, memberships); (iii) Usage-based, i.e. pay-per-use (e.g., cloud services, utilities); (iv) Licensing: fees for using intellectual property (e.g., software, patents), etc. Pricing Mechanisms should be defined here.

[7] Porter, M. E. (1985). Competitive advantage: Creating and sustaining superior performance. NY: Free Press, 1985.

[8] Tarnovskaya, V. (2023) Sustainability as the Source of Competitive Advantage. How Sustainable is it? In Creating a sustainable competitive position: Ethical Challenges for International Firms. Emerald Publishing Limited. Vol. 37.

[9] Porter, M.E. and Kramer, M.R., 2011. Creating shared value. Harvard Business Review, 89(1/2), pp. 62–77.

6. **Key Resources**—What assets are essential to deliver value to the customers and generate revenues? Usually, four groups of resources are considered: (i) Physical resources (e.g., buildings, equipment, hardware infrastructures, etc.); (ii) Intellectual resources (e.g., patents, brands, copyrights, software, etc.); (iii) Human Resources (e.g., developers, designers, scientists, sales team, leadership and strategic roles, etc.); and Financial resources (e.g., cash and lines of credit, equity, revenues, funding). This part is, thus, dedicated to define the resources needed to realize the project, deliver value to the market segments, and generate revenues. If revenues would not be sufficient to make the investment financially sustainable in a reasonable time span, it is necessary to identify other financial resources to cover the funding gap.
7. **Key Activities**—What are the most important things the company must do to make its business model work? Here the consultant helps in defining what is needed for the production activities (e.g., designing, making, and delivering a product or service, managing supply chains and logistics, quality control and continuous improvement, maintenance activities, etc.). Part of these items constitute the operating costs (OPEX) of the projects.
8. **Key Partnerships**—Who helps you deliver value? This box identifies the external companies, organizations, or individuals that help your business succeed. These partners provide resources, perform activities, or reduce risk.
9. **Cost Structure**—What are the major costs? This box outlines all the major costs involved in operating the business model. Usually costs are divided into: (i) Fixed costs (e.g. salaries, rent, insurance, etc.); (ii) Variable costs, i.e. costs that vary with production or sales volume (e.g., raw materials, energy consumption, etc.). Sometimes here there is a description of eventual economies of scale,[10] if present, or economies of scope.[11]

This canvas provides a snapshot of the business logic behind a product or service and can be represented through the Fig. 7.1.

The Original Osterwalder Business Model Canvas does not consider sustainability-linked issues. The aim of this Chapter is, therefore, to show how a traditional Business Model Canvas can be transformed in a Circular or a Sustainable Business model. In this case, exploring the general idea of adopting a more holistic view of value creation, particularly in terms of how value is generated, for whom, and how it is perceived, becomes essential.

Circularity contrasts with the traditional linear economy (take–make–dispose) by promoting a closed-loop system where materials, products, and resources are reused, repaired, refurbished, remanufactured, and recycled for as long as possible. According to the definition provided by EFRAG in the ESRS E5 Objective description, "*Circular economy means an economic system in which the value of products, materials and other resources in the economy is maintained for as long as possible,*

[10] Economies of Scale are cost advantages as production increases.

[11] Economies of scope are cost advantages from offering a variety of products/services.

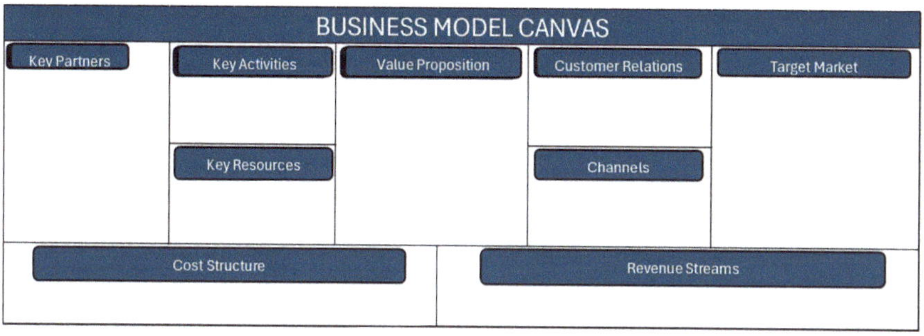

Fig. 7.1 Osterwalder Business Model Canvas Template. Source: Author elaboration on the basis of Osterwalder and Pigneur (2010) and www.strategyzer.com

enhancing their efficient use in production and consumption, thereby reducing the environmental impact of their use, minimising waste and the release of hazardous substances at all stages of their life cycle, including through the application of the waste hierarchy. The goal is to maximise and maintain the value of the technical and biological resources, products and materials by creating a system that allows for durability, optimal use or re-use, refurbishment, remanufacturing, recycling and nutrient cycling."[12]

Circular Business models (CBMs), thus, define value propositions which take into consideration how key products and materials are designed along circular principles, including *durability*, reusability, repairability, disassembly, remanufacturing, refurbishment, *recycling*, recirculation by the biological cycle, or optimization of the use of the product or material. Circular Business Models have increasingly attracted academic attention, which is unsurprising given that Circular Economy thinking fundamentally challenges the traditional linear approach to value creation. This shift underscores the importance of exploring innovative business practices that enable value creation and capture within a circular framework.[13]

It is worth to underline the positive environmental impact of the circular business modeling practice. At this regard, according to a research of Ellen MacArthur Foundation (2019),[14] implementing circular economy strategies in just five critical

[12] Source: EFRAG (2023) "ANNEX I to Commission Delegated Regulation (EU) 2023/2772 supplementing Directive 2013/34/EU of the European Parliament and of the Council as regards sustainability reporting standards, published in the Official Journal of the European Union on 22 December 2023 and including the corrigendum published on 18 April 2024.

European sustainability reporting standards (ESRS), ESRS E5.

[13] De Angelis, R. (2021). Circular economy business models: A repertoire of theoretical relationships and a research agenda. Circular Economy and Sustainability (2022) 2:433–446.

[14] Ellen MacArthur Foundation, Completing the Picture: How the Circular Economy Tackles Climate Change (2019).

sectors (cement, aluminium, steel, plastics, and food) has the potential to eliminate 9.3 billion tonnes of CO_2 emissions by 2050. This reduction is equivalent to completely eradicating current emissions from all modes of transport worldwide.

Some examples of circular business models implemented in these sectors are useful to understand how a circular business model can be designed.

In the cement sector, Geocycle[15] is a company, part of Holcim group, that uses waste materials as alternative fuels and raw materials in Holcim group cement production, reducing reliance on virgin resources and lowering CO_2 emissions.

Cemex's Regenera[16] is another project that offers circular solutions by recovering construction and demolition waste to produce alternative fuels and materials. Cemex innovated its business model by offering a wide range of tailored services including reception, management, recycling, and coprocessing of waste.

In the Plastic sector, an interesting case of circular business model is the Loop Industries[17] one. It starts using chemical recycling to break down PET plastic waste into its base monomers, which are then repolymerized into new, virgin-quality plastic.

Finally, to provide another example in the Food Industry, it is worth to mention the Protix project.[18] It converts food waste into insect protein for animal feed, closing the loop in the food production chain.

The Template to define a circular business model canvas is similar to the traditional one (See Fig. 7.1) but some questions to be answered in each building block slightly change or have to be added:[19]

1. **Customer Segments**—Who are your customers? Are there new customer segments you can target with your circular offerings?
2. **Value Propositions**—How can you maximize value creation by implementing circular strategies? E.g. increasing life span of products, offering services to maintain or repair products.
3. **Channels**—Which channels can you use to distribute our circular products and services sustainably? How can you ensure that your distribution channels support circular principles?
4. **Customer Relationships**—How can you build lasting relationships with customers by engaging them in circular initiatives?
5. **Revenue Streams**—What is the positive environmental, social, and economic impact from your companies activities to the planet (Eco-social benefits)?

---

[15] https://www.geocycle.com/

[16] https://www.cemex.com/products-solutions/regenera

[17] https://loopindustries.com/

[18] https://protix.com/

[19] Source: https://www.businessmodelhacking.com/circular-canvas-circular-business-model-canvas-template/

6. **Key Resources**—What are the key resources required to realize the circular value proposition? How can you source materials that reduce the environmental impact?
7. **Key Activities**—What key activities do you perform to deliver the circular value proposition?
8. **Key Partnerships**—How can you work with partners to create closed-loop systems?
9. **Cost Structure**—What is the negative impact on the planet from your business activities (Eco-Social Costs)?

As a matter of fact, by shifting to the concept of circular business model or to a broader concept of sustainable business models, organizations are encouraged to consider not only economic outcomes but also environmental and social impacts across the entire lifecycle of products and services.

In sustainable business model (SBM) design, two building blocks are added to the traditional business model Canvas: (i) Eco-Social Cost and (ii) Eco-Social Benefits (See Fig. 7.2).

The definition of environmental impact deals with the effects that human activities, products, and services have on the natural environment, including resource depletion, pollution, biodiversity loss, and climate change. In the context of sustainable business models, it also encompasses the systemic evaluation of these impacts across the entire lifecycle, from raw material extraction to end-of-life, aiming to minimize harm and regenerate natural systems.[20] Some of the methods used to estimate, in quantitative terms, these impacts are Life Cycle Assessment (See Sect. 6.1) and GHG Calculation

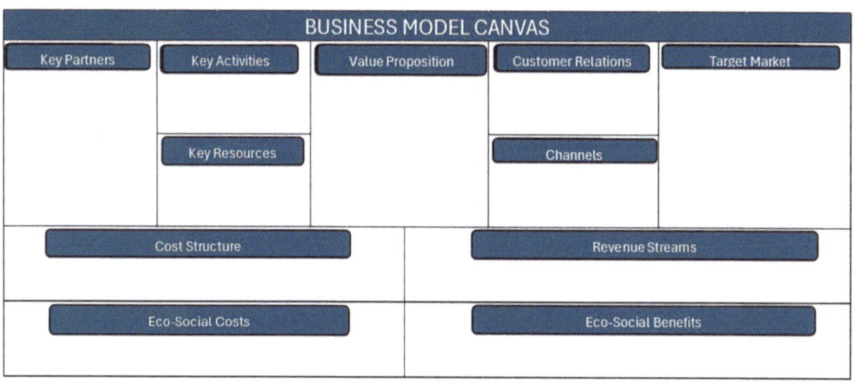

Fig. 7.2 Sustainable Business Model Canvas Template. Source: Author elaboration on the basis of www.strategyzer.com and www.case-ka.eu

[20] OECD (2025). Leveraging digital business models, tools and technologies for reliable environmental information and consumer engagement in the circular economy. OECD Digital Economy Papers, No. 377, OECD Publishing.

methods (See Sect. 6.2). They could help also in assessing circularity of the business, i.e. how and in which measure the economic model is aimed at eliminating waste and maximizing the continual use of resources. LCA, in fact, identifies where raw materials come from and where they end up, highlighting opportunities for reuse, recycling, and resource efficiency. GHG methods (e.g., GHG Protocol, ISO 14064) focus on carbon footprint but can indirectly measure circularity benefits such as Emission Reduction from Circular Practices (e.g., by substituting virgin materials with recycled ones or reducing transportation emissions through local sourcing or closed-loop systems). Combining circularity indicators and LCA/GHG ensures strategies are both resource-efficient and climate-positive, avoiding, for example, products that can be "circular" but still have high environmental impact if recycling is energy-intensive.

Regarding the definition of Social Impacts, instead, they refer to the effects that an organization's activities, products, services, or policies have on people and communities—both directly and indirectly. These impacts can be positive or negative, and they often relate to issues of equity, well-being, human rights, and social cohesion. A useful method to estimate social impacts is the Social Life Cycle Assessment (S-LCA, See ISO 14075: 2024—Environmental management—Principles and framework for social life cycle assessment). S-LCA is a methodology designed to evaluate the social and socio-economic impacts of products, services, or systems throughout their entire life cycle, from raw material extraction to disposal, similar to how environmental LCA works (see Sect. 6.1 for details), but focused on people and society.

With these premises, designing a Sustainable Business Model (SBM) may serve also as a foundational step for conducting a comprehensive cost–benefit analysis that evaluates not only the financial viability of a business or a project but also its broader environmental sustainability and impacts on society. This dual perspective encompasses both the internal viewpoint of the company and the external viewpoint of society at large, aligning with the principles of stakeholder theory and corporate responsibility (See the following Chapter for details).

7.2 From Business Planning to Cost–Benefit Analysis

Exploring the general idea of adopting a more holistic view of value creation which takes into account economic, environmental, and social costs and benefits of a project or an activity, can be translated from a qualitative approach, described in the previous Chapter regarding Sustainable Business models, to a quantitative approach through a Cost–Benefit Analysis (CBA). A Cost–Benefit Analysis (CBA) is a decision-making tool used to evaluate whether a project, investment, or policy is worthwhile by comparing its expected costs to its expected benefits, both expressed in monetary terms. Its goal is to determine if benefits outweigh costs and by how much.

The cost–benefit analysis requires an initial examination and processing of the financial data (CAPEX, OPEX and Revenues) associated with a project and then the production of a synthetic business plan (See Table 7.1 for an example).

The project illustrated in Fig. 7.1 regards example of an investment for the automation of a production line of a manufactured product. It shows a steady increase in the revenues (+5% YoY) expected from the project, over a four-year horizon, while operational expenditures remain relatively stable. As a result, EBITDA margin

Table 7.1 Example of a simplified business plan

Value in Euro	Year 1	Year 2	Year 3	Year 4
Revenue stream 1	560,000 €	588,000 €	617,400 €	648,270 €
Revenue stream 2	– €	– €	– €	– €
Total revenues	**560,000 €**	**588,000 €**	**617,400 €**	**648,270 €**
OPEX1	39,360 €	40,141 €	40,950 €	41,761 €
OPEX2	157,440 €	149,568 €	149,568 €	149,568 €
Operational expenditures (OPEX)	**196,800 €**	**189,709 €**	**190,518 €**	**191,329 €**
EBITDA	363,200 €	398,291 €	426,882 €	456,941 €
D&A	250,000 €	250,000 €	250,000 €	250,000 €
EBIT	113,200 €	148,291 €	176,882 €	206,941 €
Tax	29,432 €	38,556 €	45,989 €	53,805 €
Capital Expenditures (CAPEX)	1,000,000 €			
Net result	83,768 €	109,735 €	130,893€	153,136 €
FCF	666,232 €	359,735 €	308,893 €	403,136 €
NPV	370,098 €			

Source: Author elaboration

expands from 64.9% to 70.5%. This outcome was achieved through an efficiency-enhancing investment, requiring a capital expenditure (CAPEX) of €1 million, in the first year. Depreciation and Amortization (D&A) is fixed at €250k/year (consistent with straight-line depreciation of the €1.0m CAPEX over 4 years). The estimation of the Free cash flow (FCF) (Computed as Net Result + D&A – CAPEX, with no change in working capital) will allow the calculation of the Net Present Value (NPV) of the project and its corresponding Internal Rate of Return (IRR), necessary for the assessment of its financial sustainability. A project is generally considered financially sustainable if the Net Present Value (NPV), calculated by discounting the Free Cash Flow (FCF), is greater than zero, and the Internal Rate of Return (IRR) meets or exceeds the company's hurdle rate.

This is in fact the *conditio sine qua non* the entrepreneur is fostered to invest and therefore the environmental and social impacts of the project take place.

It is worth to underline, however, that in the example reported above, it does not emerge that the automation allows a 5% reduction in the energy consumption and thus a reduction in GHG emissions, as well as an increase in the recycling rate due to an improved categorization of wastes. The environmental benefits are not represented, as well as, possible negative externalities like a possible reduction of employees due to the automation of some processes.

In the example, we can only affirm that the project is sustainable from a financial perspective, since the Net Present Value (NPV) is positive and equal to 370 thousand euros (considering 5% discount rate). However, we cannot determine whether the broader societal benefits outweigh the associated costs.

CBA is a method used to add eco-social cost and eco-social benefits of a project from the point of view of affected entities, external stakeholders, and society at large, in order to assess whether it generates net costs or benefits and, thus, if it is sustainable from an economic point of view. The main feature of CBA is that all or most costs (or negative externalities) and benefits (new revenue/saving streams) are quantified and expressed in monetary terms.

Monetizing social costs and benefits in a Cost–Benefit Analysis (CBA) involves assigning monetary values to impacts that are not typically traded in markets—such as health, education, equity, or community cohesion. Several methods are used to assign monetary values to social impacts for example the estimates of the community willingness to pay or willingness to accept compensation for a change derived from the project development or the Social Cost of Carbon, i.e. the estimate of a monetary value for the damage caused by emitting one ton of CO_2 and so on.

Focus

Life Cycle Costing (LCC) as a Possible Component of CBA in the Monetization Phase

Life Cycle Costing (LCC) is a method for assessing the total cost of ownership of a product, system, or asset over its entire life span. This includes all costs from acquisition, operation, maintenance, and disposal. This holistic view enables decision-makers to assess not only financial implications but also environmental and social impacts, such as resource depletion, emissions, and community effects. In this term, LCC can be a component of CBA, providing detailed cost data that feeds into the broader cost–benefit framework.

According to Langdon (2014) final guidance on LCC,[21] typical benefits of carrying out a LCC study can include:

- Transparency of future operational costs;
- Ability to plan for future expenditure;
- Improved awareness of total costs;
- Ability to manipulate and optimize future costs at the design stages;
- Achieving and demonstrating better value for money in projects;
- Compliance with public sector procurement requirements;
- Evaluation of competing options, either for entire assets or parts thereof;
- Performance trade-offs against cost (e.g., environmental performance).

There are different typologies of LCC:

Conventional LCC: Focuses on direct financial costs over the life cycle of a product or project. The Environmental LCC (E-LCC) which follows the LCA system boundaries and approach (See Sect. 6.1). It uses the costs associated with the flows identified within the studied system (materials, energy services, waste production, etc.). The ideal option is to present together with LCA results.

The Societal-LCC (S-LCC) includes a larger set of costs, including those that will be, or could be, relevant in the long term for all stakeholders directly affected and for all indirectly affected through externalities. The ideal option is to present together with S-LCA results. S-LCC can be used as an input step before a full CBA because it reveals hidden or external costs and support a fair baseline for benefits quantification.

[21] Langdon, D. (2014). Life cycle costing (LCC) as a contribution to sustainable construction Guidance on the use of the LCC Methodology and its application in public procurement. European Commission, 2014.

It is worth to underline that The European Commission (EC) has developed a series of sector-specific Life-cycle costing (LCC) calculation tools which aim to facilitate the use of LCC among public procurers.[22]

Once the negative externalities/costs and benefits are monetized and discounted (if they occur over different time horizons), they are summed to costs and revenues estimated in the business plan, with the aim of identifying the Economic Net Present Value (ENPV) based on net impacts (i.e., benefits minus costs) which is useful to assess if a project is sustainable from an economic, environmental, and social point of view.

The result of the CBA is, thus, a socio-economic and environmental impact analysis and it will be summarized by the following economic and financial sustainability indicators:

- **Economic Net Present Value (ENPV)** computed by applying an appropriate discount rate to the aggregated socio-economic net benefits resulting from the implementation of the project. By treating these net benefits as a proxy for cash flows, the ENPV provides a measure of the overall value generated by the project over time.

$$\text{ENPV} = \sum_{n}^{t=0} \frac{B_t - C_t}{(1+r)^t}$$

Where:

- B_t = Benefits at time t
- C_t = Costs at time t
- r = Discount rate
- n = Time horizon

A positive ENPV would indicate that the benefits of developing the project outweigh the associated costs. ENPV, thus, indicates the sustainability of the project with respect to the financial, economic, social, and environmental aspects.

- **Economic Internal Rate of Return** which indicates the discount rate necessary to reset the ENPV (generally, the higher the IRR, the better the prospect of the investment).
- **Benefit–Cost Ratio** (B/C Ratio) according to which, the project can be considered sustainable when a value greater than 1 is obtained. When the ratio exceeds 1, it indicates that the present value of the expected benefits surpasses the present value of the associated costs. In this context, the benefit–cost ratio serves as a proxy for the project's "*Value for Money*": the higher the ratio, the greater the anticipated positive socio-economic impact per each euro invested in the project (Table 7.2).

[22] https://public-buyerscommunity.ec.europa.eu/resources/life-cycle-costing-calculation-tools-dg-env

Table 7.2 Example of simplified CBA model

Data in Euro	Year 1	Year 2	Year 3	Year 4
Revenue stream 1	560,000 €	588,000 €	617,400 €	648,270 €
Revenue stream 2	– €	– €	– €	– €
Benefit 1—GHG reduction	10,000 €	12,000 €	14,400 €	17,280 €
Benefit 2—recycling	20,000 €	24,000 €	28,000 €	34,560 €
Total revenues	**590,000 €**	**624,000 €**	**660,600 €**	**700,110 €**
opex1	39,360 €	40,141 €	40,950 €	41,761 €
Opex2	157,440 €	149,568 €	149,568 €	149,568 €
Externality1—job losses due to automation	200,000 €	200,000 €	200,000 €	200,000 €
Operational expenditure (OPEX)	**396,800 €**	**389,709 €**	**390,518 €**	**391,329 €**
EBITDA	193,200 €	234,291 €	270,082 €	308,781 €
D&A	250,000 €	250,000 €	250,000 €	250,000 €
EBIT	- 56,800 €	- 15,709 €	20,082 €	58,781 €
Tax	– €	– €	5221 €	15,283 €
CAPEX	1,000,000 €			
Net result	- 56,800 €	- 15,709 €	14,861 €	43,498 €
FCF	- 806,800 €	234,291 €	264,861 €	293,498 €
ENPV	- 89,895 €			

Source: Author elaboration

In the example above, a simplified CBA model has been presented starting from the business plan in Table 7.1. Additional costs and benefits have been identified and monetized, for instance, avoided GHG emissions, recycling rate, and expected job losses. The Economic Net Present Value (ENPV) is now negative (-89,895€) and therefore even if the project is profitable for the company from a financial point of view, negative externalities (e.g., job losses) overcome benefits to society.

The case presented above is an example to show how a project could be financially viable for the company but not convenient for the society at large. The role of the consultant is here to help companies to improve eco-social benefits and decrease or compensate eco-social costs to reach better results that could be successfully communicated to stakeholders.

To this aim, sometimes it could be useful to perform a Sensitivity Analysis. The Sensitivity Analysis aims to assess the robustness of the summary indicators, such as the Economic Net Present Value (ENPV), by identifying a range of critical variables that significantly influence the outcome of the cost–benefit assessment.

The first key output of the sensitivity analysis is the identification of these high-impact variables, upon which the viability and justification of the project most depend. This could help to prioritize interventions to improve results.

Furthermore, the results of the sensitivity analysis allow for a preliminary estimation of the overall risk associated with the project expressed as the standard deviation of the ENPV. This provides insight into the degree of uncertainty surrounding the expected net benefits and supports more informed decision-making under conditions of variability and limited predictability.

Table 7.3 Example of sensitivity analysis table

Uncertain Variables	Optimistic Estimate	Pessimistic Estimate	ENPV (Optimistic Estimate)	ENPV (Pessimistic Estimate)	Range ENPV	Sensitivity coefficient
Economic impact of Job losses	150,000.00 €	250,000.00 €	73,246.00 €	253,036.00 €	−179,790.00 €	X%
XX	XXX	XXX	XXX	XXX	XXX-XXX	X%
XX	XX	XX	XXX	XXX	XXX-XXX	X%
Most likely estimate ENPV = −0.089 million						

Notes: Range ENPV is the difference between the optimistic and the pessimistic estimates and it is an indicator of the impact of the variable on the ENPV variability. The sensitivity coefficient is given by the ratio between the variable's squared range ENPV and the sum of the total squares range ENPV. Source: Author Elaboration. Methodology from Berk et al. (2009) (Berk, J., P. DeMarzo and D. Venanzi (2009) Capital Budgeting. Pearson Paravia Bruno Mondadori)

Sensitivity checks will be, thus, conducted over key variables that significantly influence the outcome of the cost–benefit analysis. These variables will be tested across a range of plausible values—from optimistic to pessimistic scenarios—to assess the robustness of the estimated Economic Net Present Value (See Table 7.3 for an example).

In borderline cases where the project has a negative or very low ENPV even under optimistic estimates of all variables, or, on the contrary, positive values even under pessimistic estimates, the decision regarding the project's viability is straightforward.

In this context, a Monte Carlo Simulation could be also performed to simulate a simultaneous variation, between the pessimistic and the optimistic scenarios, of all the key variables, to generate a probability distribution of ENPV outcomes, providing insights into the expected average ENPV and its volatility. This probabilistic approach captures the uncertainty surrounding the assumptions and offers a more comprehensive understanding of the potential socio-economic impacts of the project.

References

Berk, J., DeMarzo, P., & Venanzi, D. (2009). *Capital budgeting*. Pearson Paravia Bruno Mondadori.

De Angelis, R. (2021). Circular economy business models: A repertoire of theoretical relationships and a research agenda. *Circular Economy and Sustainability, 2022*(2), 433–446.

Ellen MacArthur Foundation. (2019). *Completing the picture: How the circular economy tackles climate change*. Retrieved May 2021, from https://www.ellenmacarthurfoundation.org/publications/completing-the-picture-climate-change

European Commission. (2023, December 22). Commission Delegated Regulation (EU) 2023/2772 of 31 July 2023 supplementing Directive 2013/34/EU of the European Parliament and of the Council as regards sustainability reporting standards. *Official Journal of the European Union, L 322.*

Evans, S., Vladimirova, D., Holgado, M., Fossen, K. V., Yang, M., Silva, E. A., & Barlow, C. Y. (2017). Business model innovation for sustainability: Towards a unified perspective for creation of sustainable business models. *Business Strategy and the Environment, 26*, 597–608.

Johnson, E., Ramirez, C., & Oliveira, A. (2024). Integrating sustainability into corporate strategy: Challenges and opportunities. *International Journal of Management, Business, and Economics*.

Langdon, D. (2014). *Life cycle costing (LCC) as a contribution to sustainable construction guidance on the use of the LCC methodology and its application in public procurement*. European Commission.

Nguyen, H. L., & Kanbach, D. K. (2023). Toward a view of integrating corporate sustainability into strategy: A systematic literature review. *Corporate Social Responsibility and Environmental Management*.

OECD. (2025). *Leveraging digital business models, tools and technologies for reliable environmental information and consumer engagement in the circular economy* (OECD Digital Economy Papers, No. 377). OECD.

Oertwig, N., et al. (2017). Integration of sustainability into the corporate strategy. In *Sustainable manufacturing*. Springer.

Osterwalder, A., & Pigneur, Y. (2010). *Business model generation*. Wiley.

Porter, M. E. (1985). *Competitive advantage: Creating and sustaining superior performance*. Free Press.

Porter, M. E., & Kramer, M. R. (2011, January–February). Creating shared value. *Harvard Business Review, 89*(1–2), 62–77.

Spieth, P., Schneckenberg, D., & Ricart, J. E. (2014). Business model innovation—State of the art and future challenges for the field. *R&D Management, 44*(3), 237–247.

Tarnovskaya, V. (2023). Sustainability as the source of competitive advantage. How sustainable is it? In *Creating a sustainable competitive position: Ethical challenges for international firms* (Vol. 37). Emerald.

Teece, D. J. (2010). Business models, business strategy and innovation. *Long Range Planning, 43*(2/3), 172–194.

Thakkar, R. (2025, March). Integrating sustainability into corporate strategy: Challenges and best practices. *IRJAEM, 3*(03).

Zott, C., Amit, R., & Massa, L. (2011). The business model: recent developments and future research. *Journal of Management, 37*(4), 1019–1042.

Conclusions

This book has explored the evolving role of consulting in the sustainability era, emphasizing the strategic integration of environmental, social, and governance (ESG) principles into business and financial practices.

Despite substantial investments in environmentally and socially responsible products, many companies continue to see only modest financial returns and face persistent challenges in translating sustainability initiatives into profitable outcomes. This disconnect has led to growing frustration among business leaders, who struggle to align sustainability with core strategic objectives. This fact highlights the need for more integrated and adaptive business models that can bridge the gap between sustainability goals and economic performance.

Sustainability strategic consultants are increasingly seen as key actors in translating complex societal and environmental objectives into actionable business strategies which could grant an economic and financial return.

Across this book, strategic consulting emerges as a multidisciplinary practice that links strategy, finance, regulatory literacy, and impact measurement and this book offers a practical framework to support consultants in addressing sustainability-related strategic challenges through the development of customized advisory services.

The book places a strong focus on analyzing a wide range of sustainability-related services, including the support for sustainability reporting across multiple standards, climate risk assessment, and the development of resilience and adaptation action plans. The book also explores how to define a sustainable bond framework, the formulation of strategic sustainability action plans, decarbonization initiatives, and services focused on designing circular and sustainable business models and conducting cost–benefit analyses.

A common thread across all these services is the dual assessment and management of impacts: those that the organization has on the environment and society, and those that environmental and societal factors may have on the organization itself. This reciprocal perspective is encapsulated in the concept of double materiality assessment, which plays a central role in sustainability reporting.

M. Gussoni, *Integrating Strategic Sustainability Practices in Business*, CSR, Sustainability, Ethics & Governance,
https://doi.org/10.1007/978-3-032-09617-3

Another critical consideration in delivering these services is the alignment between ESG risk management strategies, sustainability investments, and corporate financial performance. Successfully integrating sustainable practices into business operations requires a holistic view of value creation. Therefore, a strategic consultant must be able to guide companies in transforming their sustainability efforts into profitable and resilient ventures.

Stakeholder engagement has also emerged as a critical success factor across all sustainability-related services. Transforming sustainability efforts into profitable investments hinges on effective communication with stakeholders. In particular, sustainable finance aims to channel capital toward businesses that embrace sustainable practices. To achieve this, investors must be adequately informed and actively involved in the strategic process. Engaging and consulting stakeholders, especially investors, is essential to shape strategies that are both impactful and financially viable.

Finally, it is important to highlight that the concepts presented in this book have become fundamental for sustainable strategic management and project development and financing. In project financing, lenders and investors rely on the project's future cash flows for repayment and sustainability analyses, covering ESG risks, regulatory compliance, and climate resilience, are critical to de-risking the project. Moreover, a large-scale project that qualify as "green" or "sustainable", because aligned with sustainability criteria, for example under Taxonomy Framework, could have an easier access to green finance instruments (e.g., green bonds) and can reduce the cost of capital. It is worth to underline, that project financing increasingly requires impact reporting to satisfy lenders, investors, and regulators. Generally speaking, access to funding remains a critical driver for business growth and, today, both private financial institutions, banks, and public funding programs require assessments that go well beyond traditional financial viability. Increasingly, these evaluations incorporate environmental, social, and governance (ESG) criteria, long-term resilience, and alignment with sustainability objectives, making integrated strategic approaches essential for securing resources and ensuring competitive advantage. For this reason, the book also serves as a valuable reference for professionals in subsidized finance, equipping them to guide companies effectively through the complexities of securing funding for sustainable projects.

MIX
Papier aus verantwortungsvollen Quellen
Paper from responsible sources
FSC® C105338

If you have any concerns about our products,
you can contact us on
ProductSafety@springernature.com

In case Publisher is established outside the EU,
the EU authorized representative is:
Springer Nature Customer Service Center GmbH
Europaplatz 3, 69115 Heidelberg, Germany

Printed by Libri Plureos GmbH
in Hamburg, Germany